MTEL General Science

10 Teacher Certification Exam

By: Sharon Wynne, M.S.
Southern Connecticut State University

XAMonline, INC.

Boston

To obtain permission(s) to use the material from this work for any purpose including workshops or seminars, please submit a written request to:

XAMonline, Inc.
21 Orient Ave.
Melrose, MA 02176
Toll Free 1-800-509-4128
Email: info@xamonline.com
Web www.xamonline.com
Fax: 1-781-662-9268

Library of Congress Cataloging-in-Publication Data

Wynne, Sharon A.
    General Science 10: Teacher Certification / Sharon A. Wynne. -2nd ed.
    ISBN 978-1-58197-593-2
    1. General Science 10.   2. Study Guides.   3. MTEL
    4. Teachers' Certification & Licensure.   5. Careers

Disclaimer:

The opinions expressed in this publication are the sole works of XAMonline and were created independently from the National Education Association, Educational Testing Service, or any State Department of Education, National Evaluation Systems or other testing affiliates.

Between the time of publication and printing, state specific standards as well as testing formats and website information may change that is not included in part or in whole within this product. Sample test questions are developed by XAMonline and reflect similar content as on real tests; however, they are not former tests. XAMonline assembles content that aligns with state standards but makes no claims nor guarantees teacher candidates a passing score. Numerical scores are determined by testing companies such as NES or ETS and then are compared with individual state standards. A passing score varies from state to state.

Printed in the United States of America   œ-1

MTEL: General Science 10
ISBN: 978-1-58197-593-2

TABLE OF CONTENTS

Great Study and Testing Tips!

What to study in order to prepare for the subject assessments is the focus of this study guide but equally important is *how* you study.

You can increase your chances of truly mastering the information by taking some simple, but effective steps.

Study Tips:

1. Some foods aid the learning process. Foods such as milk, nuts, seeds, rice, and oats help your study efforts by releasing natural memory enhancers called CCKs (*cholecystokinin*) composed of *tryptophan*, *choline*, and *phenylalanine*. All of these chemicals enhance the neurotransmitters associated with memory. Before studying, try a light, protein-rich meal of eggs, turkey, and fish. All of these foods release the memory enhancing chemicals. The better the connections, the more you comprehend.

Likewise, before you take a test, stick to a light snack of energy boosting and relaxing foods. A glass of milk, a piece of fruit, or some peanuts all release various memory-boosting chemicals and help you to relax and focus on the subject at hand.

2. Learn to take great notes. A by-product of our modern culture is that we have grown accustomed to getting our information in short doses (i.e. TV news sound bites or USA Today style newspaper articles.)

Consequently, we've subconsciously trained ourselves to assimilate information better in neat little packages. If your notes are scrawled all over the paper, it fragments the flow of the information. Strive for clarity. Newspapers use a standard format to achieve clarity. Your notes can be much clearer through use of proper formatting. A very effective format is called the *"Cornell Method."*

> Take a sheet of loose-leaf lined notebook paper and draw a line all the way down the paper about 1-2" from the left-hand edge.
>
> Draw another line across the width of the paper about 1-2" up from the bottom. Repeat this process on the reverse side of the page.

Look at the highly effective result. You have ample room for notes, a left hand margin for special emphasis items or inserting supplementary data from the textbook, a large area at the bottom for a brief summary, and a little rectangular space for just about anything you want.

3. <u>**Get the concept then the details.**</u> Too often we focus on the details and don't gather an understanding of the concept. However, if you simply memorize only dates, places, or names, you may well miss the whole point of the subject.

A key way to understand things is to put them in your own words. If you are working from a textbook, automatically summarize each paragraph in your mind. If you are outlining text, don't simply copy the author's words.

Rephrase them in your own words. You remember your own thoughts and words much better than someone else's, and subconsciously tend to associate the important details to the core concepts.

4. <u>**Ask Why?**</u> Pull apart written material paragraph by paragraph and don't forget the captions under the illustrations.

Example: If the heading is "Stream Erosion", flip it around to read "Why do streams erode?" Then answer the questions.

If you train your mind to think in a series of questions and answers, not only will you learn more, but it also helps to lessen the test anxiety because you are used to answering questions.

5. <u>**Read for reinforcement and future needs.**</u> Even if you only have 10 minutes, put your notes or a book in your hand. Your mind is similar to a computer; you have to input data in order to have it processed. *By reading, you are creating the neural connections for future retrieval.* The more times you read something, the more you reinforce the learning of ideas.

Even if you don't fully understand something on the first pass, *your mind stores much of the material for later recall.*

6. <u>**Relax to learn so go into exile.**</u> Our bodies respond to an inner clock called biorhythms. Burning the midnight oil works well for some people, but not everyone.

If possible, set aside a particular place to study that is free of distractions. Shut off the television, cell phone, pager and exile your friends and family during your study period.

If you really are bothered by silence, try background music. Light classical music at a low volume has been shown to aid in concentration over other types. Music that evokes pleasant emotions without lyrics are highly suggested. Try just about anything by Mozart. It relaxes you.

7. <u>**Use arrows not highlighters.**</u> At best, it's difficult to read a page full of yellow, pink, blue, and green streaks. Try staring at a neon sign for a while and you'll soon see that the horde of colors obscure the message.

A quick note, a brief dash of color, an underline, and an arrow pointing to a particular passage is much clearer than a horde of highlighted words.

8. <u>**Budget your study time.**</u> Although you shouldn't ignore any of the material, *allocate your available study time in the same ratio that topics may appear on the test.*

Testing Tips:

1. <u>Get smart, play dumb</u>. Don't read anything into the question. Don't make an assumption that the test writer is looking for something else than what is asked. Stick to the question as written and don't read extra things into it.

2. <u>Read the question and all the choices *twice* before answering the question</u>. You may miss something by not carefully reading, and then re-reading both the question and the answers.

If you really don't have a clue as to the right answer, leave it blank on the first time through. Go on to the other questions, as they may provide a clue as to how to answer the skipped questions.

If later on, you still can't answer the skipped ones . . . *Guess.* The only penalty for guessing is that you *might* get it wrong. Only one thing is certain; if you don't put anything down, you will get it wrong!

3. <u>Turn the question into a statement</u>. Look at the way the questions are worded. The syntax of the question usually provides a clue. Does it seem more familiar as a statement rather than as a question? Does it sound strange?

By turning a question into a statement, you may be able to spot if an answer sounds right, and it may also trigger memories of material you have read.

4. <u>Look for hidden clues</u>. It's actually very difficult to compose multiple-foil (choice) questions without giving away part of the answer in the options presented.

In most multiple-choice questions you can often readily eliminate one or two of the potential answers. This leaves you with only two real possibilities and automatically your odds go to Fifty-Fifty for very little work.

5. <u>Trust your instincts</u>. For every fact that you have read, you subconsciously retain something of that knowledge. On questions that you aren't really certain about, go with your basic instincts. **Your first impression on how to answer a question is usually correct.**

6. <u>Mark your answers directly on the test booklet</u>. Don't bother trying to fill in the optical scan sheet on the first pass through the test.

Just be very careful not to miss-mark your answers when you eventually transcribe them to the scan sheet.

7. <u>Watch the clock</u>! You have a set amount of time to answer the questions. Don't get bogged down trying to answer a single question at the expense of 10 questions you can more readily answer.

SUBAREA I.	HISTORY, PHILOSOPHY, AND METHODOLOGY OF SCIENCE

Competency 1.0   **Understand the nature of scientific thought and inquiry and the historical development of major scientific ideas**

Science is searching for information by making educated guesses and performing experiments. The experiments must be replicable. Another scientist must be able to achieve the same results under the same conditions. Science changes over time and is limited by the available technology. An example of this is the relationship of the discovery of the cell in biology and the invention of the microscope. As our technology improves, more hypotheses will become theories and possibly laws.

Ancient people believed in the geocentric theory of the solar system, which was displaced by the heliocentric theory developed by Copernicus and Kepler. Newton discovered the relationship between mass, force and acceleration, as well as the universal law of gravity. Dalton and Lavoisier made significant contributions in the fields of atom and matter.

In the 20th century, Einstein discovered relativity and the famous equation E = mc^2. The Curies and Rutherford contributed greatly to understanding radioactivity and nuclear physics. Darwin proposed his theory of evolution and Mendel's experiments on peas helped us to understand heredity. Wegener proposed his theory of continental drift, stating that continents moved away from the super continent, Pangaea.

The history of biology follows man's understanding of the living world from the earliest recorded history to modern times. Though the concept of biology as a field of science arose only in the 19th century, its origins can be traced back to the ancient Greeks (Galen and Aristotle).

Anton van Leeuwenhoek is known as the father of microscopy. In the 1650s, Leeuwenhoek began making tiny lenses that gave magnifications up to 300X. He was the first to see and describe bacteria, yeast plants, and the microscopic life found in water. Over the years, light microscopes have advanced to produce greater clarity and magnification. The transmission electron microscope (TEM) was developed in the 1950s. Instead of light, a beam of electrons passes through the specimen. Transmission electron microscopes have a resolution about one thousand times greater than light microscopes. The disadvantage of the TEM is that the chemical and physical methods used to prepare the sample result in the death of the specimen.

Carl Von Linnaeus (1707-1778), a Swedish botanist, physician, and zoologist, is well known for his contributions in ecology and taxonomy. Linnaeus is famous for his binomial system of nomenclature in which each living organism has two names, a genus and a species name. He is considered the father of modern ecology and taxonomy.

In the late 1800s, Pasteur discovered the role of microorganisms in causing disease, pasteurization, and the rabies vaccine. Koch took his observations one step further by postulating that specific diseases were caused by specific pathogens. Koch's postulates are still used as guidelines in the field of microbiology. They state that the same pathogen must be found in every diseased person, the pathogen can be isolated and grown in a culture, the disease can be induced in experimental animals from the culture, and the same pathogen can be isolated from the experimental animal.

The use of animals in biological research has expedited many scientific discoveries. Animal research has allowed scientists to learn more about animal biological systems, including the circulatory and reproductive systems. One significant use of animals is for the testing of drugs, vaccines, and other products (such as perfumes and shampoos) before use or consumption by humans. There are both significant pros and cons of animal research. The debate about the ethical treatment of animals has been ongoing since the introduction of animals to research. Many people believe the use of animals in research is cruel and unnecessary. Animal use is federally and locally regulated. The purpose of the Institutional Animal Care and Use Committee (IACUC) is to oversee and evaluate all aspects of an institution's animal care and use program.

In the 1950s, Watson and Crick discovered that the structure of a DNA molecule was that of a double helix. This structure made it possible to explain DNA's ability to replicate and to control the synthesis of proteins.

Science is a complex activity involving various people and places. A scientist may work alone or in a laboratory, classroom or for that matter, anywhere. Mostly it is a group activity requiring lot of social skills of cooperation, communication of results or findings, consultations, discussions etc. Science demands a high degree of communication to the governments, funding authorities, and to the public. The most significant result of all these discoveries was the industrial revolution in Britain, in which science was applied to increase the productivity of labor.

Competency 2.0   Understand the principles and procedures of research and experimental design

The scientific method is the basic process behind science. It involves several steps beginning with posing a question and working through to the conclusion.

Posing a question
Although many discoveries happen by chance, the standard thought process of a scientist begins with forming a question to research. The more limited the question, the easier it is to set up an experiment to answer it.

Forming a hypothesis
Once the question is formulated, take an educated guess about the answer to the problem or question. This 'best guess' is your hypothesis.

Conducting the test
To conduct an experiment there must be something that is observed or measured. An experiment observes a **variable** or any condition that can be changed, such as temperature or mass. A good test will manipulate as few variables as possible so as to see which variable is responsible for the result. Measurable quantities that remain the same in an experiment are called **controls**.

Observe and record the data
Reporting of the data should state specifics of how the measurements were calculated. A graduated cylinder, for example, needs to be read with proper procedures. For beginning students, technique must be part of the instructional process so as to give validity to the data.

Drawing a conclusion
A conclusion is the judgment derived from the data of the experiment. A laboratory report should include a specific **title** and tell what is being studied. The **abstract** is a summary of the report written at the beginning of the paper. The **purpose** of the experiment should always be defined and will state the problem. The purpose should also include the **hypothesis**. It is important to describe exactly what was done to prove or disprove a hypothesis. A **control** is necessary to prove that the results occurred from the changed variables. Only one variable should be manipulated at a time. **Observations** and **results** of the experiment should be recorded including all results from data. Drawings, graphs, and illustrations should be included to support information. Observations are objective, whereas analysis and interpretation is subjective. A **conclusion** should explain why the results of the experiment either proved or disproved the hypothesis.

Competency 3.0 **Understand the procedures for gathering, organizing, interpreting, evaluating, and communicating scientific information**

Use appropriate methods, tools, and technologies.
Experiments consist of **controls** and **variables**. A control is the experiment run under normal conditions. The variable includes a factor that is changed. In biology, the variable may be light, temperature, pH, time, etc. The differences in tested variables may be used to make a prediction or form a hypothesis. Only one variable should be tested at a time. For example, one would not alter both the temperature and pH of the experimental subject.

An **independent variable** is one that is changed or manipulated by the researcher. This could be the amount of light given to a plant or the temperature at which bacteria is grown. The **dependent variable** is that which is influenced by the independent variable.

Scientists use a variety of tools and technologies to perform tests, collect and display data, and analyze relationships. Examples of commonly used tools include computer-linked probes, spreadsheets, and graphing.

Computer-linked probes are used to measure various environmental factors including temperature, dissolved oxygen, pH, ionic concentration, and pressure. The advantage of computer-linked probes, as compared to more traditional observational tools, is that the probes automatically gather data and present it in an accessible format.

Spreadsheets are used to organize, analyze, and display data. For example, conservation ecologists use spreadsheets to model population growth and development, apply sampling techniques, and create statistical distributions to analyze relationships. Spreadsheet use simplifies data collection and manipulation and allows the presentation of data in a logical and understandable format.

Graphing calculators are another technology with many applications to science. For example, biologists use algebraic functions to analyze growth, development and other natural processes. Graphing calculators can manipulate algebraic data and create graphs for analysis and observation. In addition, biologists use the matrix function of graphing calculators to model problems in genetics. The use of graphing calculators simplifies the creation of graphical displays including histograms, scatter plots, and line graphs.

Select appropriate methods and criteria for organizing and displaying data.

The type of graphic representation used to display observations depends on the data that is collected. **Line graphs** are used to compare different sets of related data or to predict data that has not yet been measured. An example of a line graph would be comparing the rate of activity of different enzymes at varying temperatures. A **bar graph** or **histogram** is used to compare different items and make comparisons based on this data. An example of a bar graph would be comparing the ages of children in a classroom. A **pie chart** is useful when organizing data as part of a whole. A good use for a pie chart would be displaying the percent of time students spend on various after school activities.

As noted before, the independent variable is controlled by the experimenter. This variable is placed on the x-axis (horizontal axis). The dependent variable is influenced by the independent variable and is placed on the y-axis (vertical axis). It is important to choose the appropriate units for labeling the axes. It is best to take the largest value to be plotted and divide it by the number of blocks, and rounding to the nearest whole number.

Demonstrate an understanding of the concepts of precision, accuracy, and error.

Accuracy is the degree of conformity of a measured, calculated quantity to its actual (true) value. Precision also called reproducibility or repeatability and is the degree to which further measurements or calculations will show the same or similar results.

Accuracy is the degree of veracity while precision is the degree of reproducibility. Repeated measurements can be compared to arrows that are fired at a target. Accuracy describes the closeness of arrows to the bull's eye at the target center. Arrows that strike closer to the bull's eye are considered more accurate. While precision is how many arrows cluster together on the target irregardless whether they strike the target.

All experimental uncertainty is due to either random errors or systematic errors.

Random errors are statistical fluctuations in the measured data due to the precision limitations of the measurement device. Random errors usually result from the experimenter's inability to take the same measurement in exactly the same way to get exactly the same number.

Systematic errors, by contrast, are reproducible inaccuracies that are consistently in the same direction. Systematic errors are often due to a problem that persists throughout the entire experiment.

Systematic and random errors refer to problems associated with making measurements. Mistakes made in the calculations or in reading the instrument are not considered in error analysis.

Identify and evaluate various sources of scientific information

Because people often attempt to use scientific evidence in support of political or personal agendas, the ability to evaluate the credibility of scientific claims is a necessary skill in today's society. In evaluating scientific claims made in the media, public debates, and advertising, one should follow several guidelines.

First, scientific, peer-reviewed journals are the most accepted source for information on scientific experiments and studies. One should carefully scrutinize any claim that does not reference peer-reviewed literature.

Second, the media and those with an agenda to advance (advertisers, debaters, etc.) often overemphasize the certainty and importance of experimental results. One should question any scientific claim that sounds fantastical or overly certain.

Finally, knowledge of experimental design and the scientific method is important in evaluating the credibility of studies. For example, one should look for the inclusion of control groups and the presence of data to support the given conclusions.

Competency 4.0   **Understand the safe and proper use of tools, equipment, and materials (including chemicals and living organisms) related to classroom and other science investigations**

Hot plates - Hot plates, not Bunsen burners, should be used whenever possible to avoid the risk of burns or fire.

Graduated Cylinder - These are used for precise measurements. They should always be placed on a flat surface. The surface of the liquid will form a meniscus (lens-shaped curve). The measurement is read at the <u>bottom</u> of this curve.

Balance - An electronic balance should always be tared (returned to zero) before measuring and used on a flat surface. Substances should always be placed on a piece of paper to avoid spills and/or damage to the instrument. Triple beam balances must be used on a level surface. There are screws located at the bottom of the balance to make any adjustments.

Buret – A buret is used to dispense precisely measured volumes of liquid. A stopcock is used to control the volume of liquid being dispensed at a time.

Light microscopes - Microscopes are commonly used in laboratory experiments. Several procedures should be followed to properly care for this equipment:

- Clean all lenses with lens paper only.
- Carry microscopes with two hands; one on the arm and one on the base.
- Always begin focusing on low power, then switch to high power.
- Store microscopes with the low power objective down.
- Always use a coverslip when viewing wet mount slides.
- Bring the objective down to its lowest position then focus by moving up to avoid breaking the slide or scratching the lens.

Centrifuge - A centrifuge involves spins substances at a high speed. The more dense part of a solution will settle to the bottom of the test tube, while the lighter material will stay on top. Centrifugation is used to separate blood into blood cells and plasma, with the heavier blood cells settling to the bottom.

Safe practices and procedures in all areas related to science instruction

In addition to requirements set forth by your place of employment, the NABT (National Association of Biology Teachers) and ISEF (International Science Education Foundation) have been instrumental in setting parameters for the science classroom. All science labs should contain the following items of **safety equipment**. Those marked with an asterisk are requirements by state laws.

* fire blanket which is visible and accessible
*Ground Fault Circuit Interrupters (GCFI) within two feet of water supplies
*signs designating room exits
*emergency shower providing a continuous flow of water
*emergency eye wash station which can be activated by the foot or forearm
*eye protection for every student and a means of sanitizing equipment
*emergency exhaust fans providing ventilation to the outside of the building
*master cut-off switches for gas, electric and compressed air. Switches must have permanently attached handles. Cut-off switches must be clearly labeled.
*an ABC fire extinguisher
*storage cabinets for flammable materials
-chemical spill control kit
-fume hood with a motor which is spark proof
-protective laboratory aprons made of flame retardant material
-signs which will alert potential hazardous conditions
-containers for broken glassware, flammables, corrosives, and waste
-containers should be labeled.

Students should wear safety goggles when performing dissections, heating, or while using acids and bases. Hair should always be tied back and objects should never be placed in the mouth. Food should not be consumed while in the laboratory. Hands should always be washed before and after laboratory experiments. In case of an accident, eye washes and showers should be used for eye contamination or a chemical spill that covers the student's body. Small chemical spills should only be contained and cleaned by the teacher. Kitty litter or a chemical spill kit should be used to clean spill. For large spills, the school administration and the local fire department should be notified. Biological spills should also be handled only by the teacher. Contamination with biological waste can be cleaned by using bleach when appropriate.

Accidents and injuries should always be reported to the school administration and local health facilities. The severity of the accident or injury will determine the course of action to pursue.

It is the responsibility of the teacher to provide a safe environment for their students. Proper supervision greatly reduces the risk of injury and a teacher should never leave a class for any reason without providing alternate supervision. After an accident, two factors are considered; **foreseeability** and **negligence**. Foreseeability means the teacher could have reasonably foreseen that an event may occur under certain circumstances. Negligence is the failure to exercise ordinary or reasonable care. Safety procedures should be a part of the science curriculum and a well-managed classroom is important to avoid potential lawsuits.

First response procedures, including first aid, for responding to accidents

All staff should be trained in first aid in the science classroom and laboratory. Please remember to always report all accidents, however minor, to the lab instructor immediately. In most situations 911 should immediately be called. Please refer to your school's specific safety plan for accidents in the classroom and laboratory. The classroom/laboratory should have a complete first-aid kit with supplies that are up-to-date and checked frequently for expiration.

Know the location and use of fire extinguishers, eye-wash stations, and safety showers in the lab.

Do not attempt to smother a fire in a beaker or flask with a fire extinguisher. The force of the stream of material from it will turn over the vessel and result in a bigger fire. Just place a watch glass or a wet towel over the container to cut off the supply of oxygen.

If your clothing is on fire, **do not run** because this only increases the burning. It is normally best to fall on the floor and roll over to smother the fire. If a student, whose clothing is on fire panics and begins to run, attempt to get the student on the floor and roll over to smother the flame. If necessary, use the fire blanket or safety shower in the lab to smother the fire.

Students with long hair should put their hair in a bun or a pony-tail to avoid their hair catching fire.

Below are common accidents that everyone who uses the laboratory should be trained in how to respond:

Burns (Chemical or Fire) – Use deluge shower for 15 minutes.

Burns (Clothing on fire) – Use safety shower immediately. Keep victim immersed 15 minutes to wash away both heat and chemicals. All burns should be examined by medical personnel.

Chemical spills – Chemical spills on hands or arms should be washed immediately with soap and water. Washing hands should become an instinctive response to any chemical spilled on hands. Spills that cover clothing and other parts of the body should be drenched under the safety shower. If strong acids or bases are spilled on clothing, the clothing should be removed. If a large area is affected, remove clothing and immerse victim in the safety shower. If a small area is affected, remove article of clothing and use a deluge shower for 15 minutes.

Eyes (chemical contamination) – Hold the eye wide open and flush with water from the eye wash for about 15 minutes. Seek medical attention.

Ingestion of chemicals or poisoning – See antidote chart on wall of lab for general first-aid directions. The victim should drink large amounts of water. All chemical poisonings should receive medical attention.

Information about safety, legal issues, and the proper use, storage, and proper disposal of scientific materials

The 'Right to Know Law' covers science teachers who work with potentially hazardous chemicals. Briefly, the law states that employees must be informed of potentially toxic chemicals. An inventory must be made available if requested. The inventory must contain information about the hazards and properties of the chemicals. This inventory is to be checked against the 'Substance List.' Training must be provided on the safe handling and interpretation of the Material Safety Data Sheet.

The following chemicals are potential carcinogens and not allowed in school facilities: acrylonitriel, arsenic compounds, asbestos, bensidine, benzene, cadmium compounds, chloroform, chromium compounds, ethylene oxide, ortho-toluidine, nickel powder, and mercury.

Chemicals should not be stored on bench tops or heat sources. They should be stored in groups based on their reactivity with one another and in protective storage cabinets. All containers within the lab must be labeled. Suspected and known carcinogens must be labeled as such and segregated within trays to contain leaks and spills.

Chemical waste should be disposed of in properly labeled containers. Waste should be separated based on their reactivity with other chemicals. Biological material should never be stored near food or water used for human consumption. All biological material should be appropriately labeled. All blood and body fluids should be put in a well-contained container with a secure lid to prevent leaking. All biological waste should be disposed of in biological hazardous waste bags.

As a teacher, you should utilize a MSDS (Material Safety Data Sheet) whenever you are preparing an experiment. It is designed to provide people with the proper procedures for handling or working with a particular substance. An MSDS includes information such as physical data (melting point, boiling point, etc.), toxicity, health effects, first aid, reactivity, storage, disposal, protective gear, and spill/leak procedures. These are particularly important if a spill or other accident occurs. You should review a few, available commonly online, and understand the listing procedures. Material safety data sheets are available directly from the company of acquisition or the internet. The manuals for equipment used in the lab should be read and understood before using them.

Use and care of living organisms in an ethical and appropriate manner

No dissections may be performed on living mammalian vertebrates or birds. Lower order life and invertebrates may be used. Biological experiments may be done with all animals except mammalian vertebrates or birds. No physiological harm may result to the animal. All animals housed and cared for in the school must be handled in a safe and humane manner. Animals are not to remain on school premises during extended vacations unless adequate care is provided. Any instructor who intentionally refuses to comply with the laws may be suspended or dismissed.

Pathogenic organisms must never be used for experimentation. Students should adhere to the following rules at all times when working with microorganisms to avoid accidental contamination:

1. Treat all microorganisms as if they were pathogenic.
2. Maintain sterile conditions at all times

SUBAREA II.   **CHEMISTRY**

Competency 5.0   **Understand the structure and nature of matter**

Everything in our world is made up of **matter**, whether it is a rock, a building, or an animal. **Mass** is a measure of the amount of matter in an object. Two objects of equal mass will balance each other on a simple balance scale and the units of mass are kilograms in the metric system of units.

All matter is made up of 109 **elements** or **atoms** which are listed on the periodic table. The following is the entry for the element carbon:

Element Key

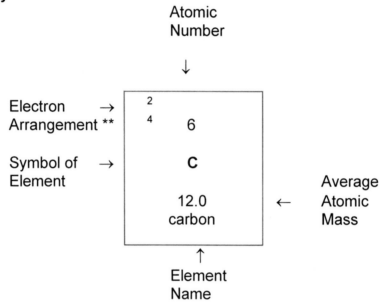

** Number of electrons on each level. Top number represents the innermost level.

The **nucleus** is the center of the atom. The positive particles inside the nucleus are called **protons.** The mass of a proton is about 2,000 times that of the mass of an electron. The number of protons in the nucleus of an atom is called the **atomic number**. All atoms of the same element have the same atomic number.

Neutrons are another type of particle in the nucleus. Neutrons and protons have about the same mass, but neutrons have no charge. Neutrons were discovered because scientists observed that not all atoms in neon gas have the same mass. **Isotopes** of an element have the same number of protons in the nucleus, but have different masses.

Scientists measure the mass of an atom by comparing it to that of a standard atom which is an isotope of the element carbon. It has six (6) neutrons and is called carbon-12. It is assigned a mass of 12 atomic mass units (amu). Therefore, the **atomic mass unit (amu)** is the standard unit for measuring the mass of atoms.

The **mass number** of an atom is the sum of its protons and neutrons. In any element, there is a mixture of isotopes, some having slightly more or slightly fewer protons and neutrons. The **atomic mass** of an element is an average of the mass numbers of its atoms.

Each atom has an equal number of electrons (negative) and protons (positive). Electrons orbiting the nucleus occupy energy levels that are arranged in order and the electrons tend to occupy the lowest energy level available. A **stable electron arrangement** is an atom that has all of its electrons in the lowest possible energy levels.

Each energy level holds a maximum number of electrons. However, an atom with more than one level does not hold more than 8 electrons in its outermost shell.

Level	Name	Max. # of Electrons
First	K shell	2
Second	L shell	8
Third	M shell	18
Fourth	N shell	32

This explains why chemical reactions occur. Atoms react with each other when their outer energy levels are unfilled. When atoms either exchange or share electrons with each other, these energy levels become filled and the combination becomes more stable.

A **compound** is made of two or more elements that have been chemically combined. Atoms join together when elements are chemically combined. The result is that the elements lose their individual identities when they are joined. The compound that they become has different properties.

A **chemical formula** is a shorthand way of showing what elements are in a compound. The letter symbols are taken from the periodic table and the subscripts tells how many atoms of each element are in the compound. No subscript is used if there is only one atom. For example, carbon dioxide is made up of one atom of carbon (C) and two atoms of oxygen (O_2), so the formula would be represented as CO_2.

The **periodic table of elements** is an arrangement of the elements in rows and columns so that it is easy to locate elements with similar properties. The elements of the modern periodic table are arranged in numerical order by atomic number.

The **periods** are the rows down the left side of the table. They are called first period, second period, etc. The columns of the periodic table are called **groups**, or **families.** Elements in a family have similar properties.

The periodic table arranges metals into families with similar properties. The periodic table has its columns marked IA–VIIIA. These are the traditional group numbers. Arabic numbers 1–18 are also used, as suggested by the Union of Physicists and Chemists.

With the exception of hydrogen, all elements in Group 1 are **alkali metals**. These metals are shiny, softer and less dense than other metals, and are the most chemically active.

Group 2 metals are the **alkaline earth metals.** They are harder, denser, have higher melting points, and are chemically active.

The **transition elements** can be found by finding the periods (rows) from 4 to 7 under the groups (columns) 3 - 12. They are metals that do not show a range of properties as you move across the chart. They are hard and have high melting points. Compounds of these elements are colorful, such as silver, gold, and mercury.

Nonmetals are not as easy to recognize as metals because they do not always share physical properties. However, in general the properties of nonmetals are the opposite of metals. They are dull, brittle, and are not good conductors of heat and electricity. Nonmetals include solids, gases, and one liquid (bromine).

The **halogens** can be found in Group 17. Halogens combine readily with metals to form salts. Table salt, fluoride toothpaste, and bleach all have an element from the halogen family.

The **noble gases** got their name from the fact that they did not react chemically with other elements, much like the nobility did not mix with the masses. These gases (found in Group 18) will only combine with other elements under very specific conditions. They are **inert** (inactive).

Metalloids have properties in between metals and nonmetals. They can be found in Groups 13–16, but do not occupy the entire group. They are arranged in stair steps across the groups.

Competency 6.0   Understand the nature of physical changes in matter

Common compounds are **acids, bases, salts**, and **oxides** and are classified according to their characteristics.

An **acid** contains at least one hydrogen (H) atom. Although it is never wise to taste a substance to identify it, acids have a sour taste. Vinegar and lemon juice are both acids, and acids occur in many foods in a weak state. Strong acids can burn skin and destroy materials. Common acids include:

Sulfuric acid (H_2SO_4)	-	Used in medicines, alcohol, dyes, and car batteries
Nitric acid (HNO_3)	-	Used in fertilizers, explosives, cleaning materials
Carbonic acid (H_2CO_3)	-	Used in soft drinks
Acetic acid ($HC_2H_3O_2$)	-	Used in making plastics, rubber, photographic film, and as a solvent

Bases have a bitter taste and the stronger ones feel slippery. Like acids, strong bases can be dangerous and should be handled carefully. All bases contain the elements oxygen and hydrogen (OH). Many household cleaning products contain bases. Common bases include:

Sodium hydroxide	NaOH	-	Used in making soap, paper, vegetable oils, and refining petroleum
Ammonium hydroxide	NH_4OH	-	Making deodorants, bleaching compounds, cleaning compounds
Potassium hydroxide	KOH	-	Making soaps, drugs, dyes, alkaline batteries, and purifying industrial gases
Calcium hydroxide	$Ca(OH)_2$	-	Making cement and plaster

Salt is formed when an acid and a base combine chemically. Water is also formed. The process is called **neutralization**. Table salt (NaCl) is an example of this process. Salts are also used in toothpaste, Epsom salts, and cream of tartar. Calcium chloride ($CaCl_2$) is used on frozen streets and walkways to melt the ice.

Oxides are compounds that are formed when oxygen combines with another element. Rust is an oxide formed when oxygen combines with iron.

Matter constantly changes. A **physical change** is a change that does not produce a new substance. The freezing and melting of water is an example of physical change. A **chemical change** (or chemical reaction) is any change of a substance into one or more other substances. Burning materials turn into smoke; a seltzer tablet fizzes into gas bubbles.

The **phase of matter** (solid, liquid, or gas) is identified by its shape and volume. A **solid** has a definite shape and volume. A **liquid** has a definite volume, but keeps the shape of its container. A **gas** has no shape or volume because it will spread out to occupy the entire space of whatever container it is in. Applying heat to a frozen liquid changes it from solid back to liquid. Continue heating it and it will boil and give off steam, a gas. Changes in phase are considered physical changes. **Evaporation** is the change in phase from liquid to gas. **Condensation** is the change in phase from gas to liquid.

Substances can combine without a chemical change. A **mixture** is any combination of two or more substances in which the substances keep their own properties. A fruit salad is a mixture.

Another example of a mixture is a **solution**. The minor component of a solution is called the **solute** and the major component is called the **solvent**. Water and oil or water and sand do not mix, however, salt and sugar dissolve when mixed with water.

The chlorine and sodium atoms in salt are connected by an ionic bond which means an electron transfers from the sodium atom to the chlorine atom. In water, the sodium and chlorine separate and become sodium ions and chlorine ions. The formula for sugar is $C_6H_{12}O_6$ and the atoms are bonded by sharing electrons. This type of bond is called a covalent bond and the sugar molecules do not dissociate.

Competency 7.0   Understand the nature of chemical changes in matter

One or more substances are formed during a **chemical reaction**. In **exothermic** reactions energy is released and in **endothermic** reactions energy is absorbed from the environment. In a fireworks display or fire, energy is released very rapidly. However, the chemical reaction that produces tarnish on a silver spoon happens very slowly.

Chemical equilibrium is defined as occurring when the quantities of reactants and products do not change. The rate of forward reaction must equal the rate of backward reaction.

In a chemical reaction, the total mass of the reactants is equal to the total mass of the products. This is called the conservation of mass. This principle is used to balance the equations used to describe chemical reactions. For example, carbon and oxygen react to form carbon dioxide. The balanced chemical equation can be written:

$$C \quad + \quad O_2 \quad \rightarrow \quad CO_2$$

1 atom of carbon	+	2 atoms of oxygen	→ →	1 molecule of carbon dioxide

An electric current can split water molecules into hydrogen and oxygen gas:

$$2H_2O \quad \rightarrow \quad 2H_2 \quad + \quad O_2$$

2 molecules of water	→	2 molecules of hydrogen	+	1 molecule of oxygen

An example of one element taking the place of another is when iron changes places with copper in the compound copper sulfate:

$$CuSo_4 \quad + \quad Fe \quad \rightarrow \quad FeSO_4 \quad + \quad Cu$$

copper sulfate	+	iron (steel wool)	iron sulfate	copper

Sometimes two sets of elements change places. In this example, an acid and a base are combined:

$$HCl \quad + \quad NaOH \quad \rightarrow \quad NaCl \quad + \quad H_2O$$

hydrochloric acid	+	sodium hydroxide	sodium chloride (table salt)	water

The above examples illustrate four kinds of chemical reactions. In a **composition reaction**, two or more substances combine to form a compound:

$A + B \rightarrow AB$

In a **decomposition reaction**, a compound breaks down into two or more simpler substances:

$AB \rightarrow A + B$

In a **single replacement reaction**, a free element replaces an element that is part of a compound.

$A + BX \rightarrow AX + B$

In a **double replacement reaction**, parts of two compounds replace each other. In this case, the compounds seem to switch partners.

$AX + BY \rightarrow AY + BX$

Competency 8.0   Understand the kinetic molecular model of matter.

According to the kinetic theory of matter, a substance is made up of molecules. In a solid the molecules are bound together by strong forces so that the molecules only vibrate back and forth. In a liquid the forces between the molecules are less. In a gas, the forces are so small that each molecule moves separately and a container is needed to hold the gas.

As a gas in a container is heated, the molecules begin moving faster within the container. **Pressure** (P) is the force exerted by a gas on a surface divided by the area of the surface. Pressure is measured in a unit called the pascal. One pascal (Pa) is equal to one newton of force pushing on one square meter of area. **Temperature** (T) is a property of a substance and is measured with a thermometer in units in Celsius or Fahrenheit units. The **volume** (V) of a gas is the volume of the container and is a variable. Another variable is the amount or mass of the gas which is measured in **moles**. A mole is 6.022×10^{23} molecules of the gas. This quantity is known as Avogadro's number.

At a constant pressure and mass, an increase in temperature of a gas causes an increase in the volume of a gas. This is called **Charles' Law**. At a constant temperature and mass, a decrease in the volume of a gas causes an increase in its pressure. This relationship between pressure and volume is called **Boyle's Law**. These laws can be combined to give the ideal gas law: $PV = nRT$, where n is the number of moles of the gas and R is the universal gas constant.

The universal gas constant is directly related to Boltzmann's constant, which connects the average kinetic energy of a gas molecule with the temperature of the gas. The lowest possible temperature of a gas is about $-273°C$ or 0 K on the Kelvin scale.

SUBAREA III.   PHYSICS

Competency 9.0   **Understand the concepts of force, motion, work, and power**

The concept of force comes from the experience of pushing and pulling and is clearly related to the motion of objects. It is a vector quantity because it has both direction and magnitude. Mass is a scalar quantity. The most common forces we experience are the force of gravity, elastic forces from springs, and friction.

Newton's first law is also called the law of inertia. It states that an object at rest will remain at rest and an object in motion will remain in motion at a constant velocity unless acted upon by an external force.

Newton's second law states that if a net force acts on an object, it will cause the acceleration of the object. The relationship between force and motion is that force equals mass times acceleration. (*F = ma*). The acceleration of an object is the slope of the speed-time graph of the object's motion or the change is speed divided by the time it took for the speed to change. The speed (*v*) in turn is the slope of the distance-time graph of the object's motion.

Newton's third law states that forces come in equal and opposite pairs: If an object exerts a force on another object, that second object exerts an equal and opposite force on the first.

Work is done on an object when an applied force acts through a distance. The unit of work is the joule = 1 newton × meter/(second)^2. **Power** is the work done divided by the amount of time that it took to do it, or the rate of doing work. The units of power are the watt and horsepower.

 Energy is defined as the ability to do work. The concept of work enables us to understand why simple machines, such as levers, ramps, and pulleys, can make a job easier. A simple machine transforms a short distance and large force to a long distance and a small force.

The force of gravity comes from fact that all objects with mass are attracted to one another in accordance with the universal law of gravity:

$$F_{gravity} = G\frac{m_1 m_2}{d^2}$$

G is the universal gravitational constant and can be measured in a laboratory with a torsion balance. When the measurement was first performed there was no knowledge of the mass of the earth. However, it was known that all objects at the surface of the earth accelerate at the same rate ($g = 9.8$ m/s^2). Thus *G* and the mass of the earth could be measured. Critical to these calculations is that gravitational mass is equal to the inertial mass of Newton's second law. The force of gravity is called weight.

The momentum of an object is its velocity times its mass. Velocity is a vector quantity since in includes the direction of motion as well as the speed of the object. Because of Newton's third law, it is a strict law of physics that total momentum is conserved. In a collision, the total momentum before the collision is equal to the total momentum after the collision.

Surfaces that touch each other have a certain resistance to motion. This resistance is **friction**. The following observations can be made:

1. The materials that make up the surfaces will determine the magnitude of the frictional force.
2. The frictional force is independent of the area of contact between the two surfaces.
3. The direction of the frictional force is opposite to the direction of motion.
4. When an object is resting on a surface, the frictional force is proportional to the normal force between the surface and the object.

Static friction describes the force of friction of two surfaces that are in contact but do not have any motion relative to each other, such as a block sitting on an inclined plane. **Kinetic friction** describes the force of friction of two surfaces in contact with each other when there is relative motion between the surfaces.

When an object moves in a circular path, a force must be directed toward the center of the circle in order to keep the motion going. This constraining force is called **centripetal force**. Gravity is the centripetal force that keeps a satellite circling the earth.

Competency 10.0   Understand the concept of energy and the forms that energy can take

Energy is the ability to do work. Kinetic energy is the energy of motion and potential energy is the energy of position. A coiled spring or a boulder on top of a hill has potential energy. Mechanical energy is the sum of kinetic and potential energy and is always conserved if the force depends only on position. Such forces are called conservative forces. Friction is not a conservative force and mechanical energy is obviously lost when there is friction. In elastic collisions, mechanical energy is conserved. In inelastic collisions, mechanical energy is lost.

However, there is another kind of energy called thermal energy or internal energy. For an ideal gas, the internal energy is equal to the sum of the kinetic energies of the molecules in the gas. In the case of friction, mechanical energy is transformed into internal energy.

The first law of thermodynamics states that energy is never lost or gained, it just transforms from one type of energy into another. The change in thermal energy (U) of a volume of gas is equal to the heat added to the gas minus the work done by the gas: $\Delta U = Q - W$. One can picture gas in a container with a movable piston that can be heated or cooled. However, the law is true for all possible systems.

The definition of heat is $Q = mc\Delta T$. where, m is the amount of substance, c is the specific heat of the substance, and ΔT is the change in temperature of the substance. A calorie is the amount of heat added to 1 gram of water that causes the temperature of the water to increase by 1°C. The specific heat is measured with respect to that of water, which has a higher heat capacity than any other substance. 1 calorie of heat will raise the temperature of 1 gram lead by 19°C.

Experiments show that 1 calorie = 4.186 joules. The caloric value of food is determined by burning the food in a calorimeter and is given in terms of kilocalories.

The second law of thermodynamics has a number of formulations:

1. No machine that transforms heat into work is 100% efficient because some heat will always be lost to the environment.
2. Heat cannot spontaneously pass from a colder object to a hotter object.
3. There is a tendency in nature for **entropy** to increase.

Entropy is the measure of disorder in a system. An example of entropy increasing is the expansion of a gas when its container increases in volume. After the expansion, there is more disorder because there is less knowledge about the location of the molecules. Likewise, when heat flows from a hot object to a cold object, there is less knowledge about the kinetic energy of the molecules in both objects.

When an object undergoes a change of phase, it goes from one physical state (solid, liquid, or gas) to another. For instance, water can go from liquid to solid (freezing) or from liquid to gas (boiling). The heat that is required to change from one state to the other is called **latent heat.** The **heat of fusion** is the amount of heat that it takes to change from a solid to a liquid or the amount of heat released during the change from liquid to solid. The **heat of vaporization** is the amount of heat that it takes to change from a liquid to a gaseous state.

Heat is transferred in three ways: **conduction, convection,** and **radiation.** Conduction occurs when heat travels through the heated solid. Convection is heat transported by the movement of a heated substance. Radiation is heat transfer as the result of electromagnetic waves.

An example of all three methods of heat transfer occurs in the thermos bottle or Dewar flask. The bottle is constructed of double walls of Pyrex glass that have a space in between. Air is evacuated from the space between the walls and the inner wall is silvered. The lack of air between the walls lessens heat loss by convection and conduction. The heat inside is reflected by the silver, cutting down heat transfer by radiation. Hot liquids remain hotter and cold liquids remain colder for longer periods of time.

Competency 11.0   Understand characteristics of waves and the behavior of sound and light waves

Sound waves, water waves, **transverse** waves on a string, and **longitudinal** waves in a spring are examples of mechanical waves. In mechanical waves there is a transfer of energy without any bulk motion of the medium the wave propagates in. A wave is a series of pulses produced by some kind of vibration.

The **period** of a wave is the time between pulses and the **frequency** is the number of pulses per second. The unit of frequency is the hertz. The **amplitude** of the wave is how much the medium is displaced and the **speed** of the wave is how fast the pulses travel. The **wavelength** is the distance between pulses.

Sound is a longitudinal wave because the molecules of the solid, liquid, or gas vibrate back and forth in the same direction that energy is propagated. Our sense of hearing gives us the sensation of loudness and pitch when sound travels in air. Loudness comes from the amplitude of the sound wave and pitch comes from the frequency of the sound wave.

Interference is the interaction of two or more waves traveling in the same medium at the same time. If the waves interfere constructively, the amplitudes become larger. If the waves interfere destructively, the amplitudes decrease or cancel out. When you simultaneously strike the two tuning forks with frequencies that are similar, you may hear **beats**. Beats are a series of loud and soft sounds. This phenomena occurs because when the waves meet, the crests combine at some points and produce louder sounds. At other points, they nearly cancel each other out and produce softer sounds.

Another property of mechanical waves is that they **diffract**, **refract**, and **reflect**. Diffraction means they bend around objects in its path and generally spread out in the medium. Refraction and reflection occur when a wave's medium changes. The reflected wave bounces of the interface and the refracted wave is propagated through the new medium.

Electromagnetic radiation includes radio waves, microwaves, infrared radiation, visible light, ultraviolet radiation, X-rays, and gamma rays. Electromagnetic radiation travels in a vacuum, so it is not a mechanical wave. It consists of light particles called photons. The photons have no mass, but have energy, a wavelength, and a frequency. In a vacuum light travels at a speed of 186,000 miles per second. It travels slower in transparent mediums, such as water and glass. Visible light appears to travel in a straight line, but a closer look at the edge of shadows shows there is some diffraction. Radio wave obviously diffract around buildings and mountains.

You can observe the diffraction and interference of light by making a very thin slit between your thumb and forefinger. Hold them about 8 cm from your eye and look at a distant source of light. The pattern you observe is caused by the diffraction of light.

Optics is the study of light rays which are assumed to travel in a straight line unless they strike glass, or some other transparent medium, at an angle. In this case, part of the ray reflects and the other part refracts. The angle of refraction depends on the properties of the transparent medium and is governed by Snell's law. Light traveling from a transparent medium to air may not be refracted at all. This is called total internal reflection and is the basis of fiber optics.

The images created by convex and concave mirrors and lenses can be understood from optics. The image that you see in a bathroom mirror is a virtual image because it only seems to be where it is. No light rays go through the image point. However, a curved mirror can produce a real image. A real image is produced when light passes through the point where the image appears. A real image can be projected onto a screen.

Competency 12.0 Understand principles of electricity, magnetism, and electromagnetism

A plastic rod that is rubbed with fur and a glass rod that is rubbed with silk will become electrically charged. The charge on the plastic rod rubbed with fur is negative and the charge on glass rod rubbed with silk is positive. Unlike charges attract each other with a force that follows an inverse square law similar to the universal law of gravity. Like charges repel each other. Objects acquire a negative charge when they gain electrons and acquire a positive charge when they lose electrons. Charge is measured in units called coulombs.

Lodestones are found in nature and possess magnetic properties. There are no magnetic charges because a magnet always has a north pole and south pole. You cannot separate the two poles, but like poles repel and unlike poles attract, as with electrical charges.

You can give a charge to a piece of metal by placing it in water. The water molecules pull metallic ions out of the metal into the water, leaving behind electrons. This gives the metal electrode, as it is called, a negative charge. Copper produces a bigger negative charge than zinc, which is the basis of a zinc-copper battery. In metal, the electrons are not bound to any particular atom and are free to move. Electrons will flow in a wire when the wire is connected to the electrodes of a battery because the batter produces an **electric field** in the wire. A stationery electric charge can be thought of as producing an electric field that radiates outward in all directions with a magnitude that decreases according to the inverse square law.

Moving electrons produce circular **magnetic fields**. What is more, moving electric fields produce magnetic fields and moving magnetic fields produce electric fields. The behavior of magnetic and electrical forces and fields are governed by a set of laws known as Maxwell's equations. The speed of light can be derived from Maxwell's equations, which shows that electromagnetic radiation consists of electric and magnetic fields propagating through space.

An **electric circuit** is a path along which electrons flow. A simple circuit can be created with a battery, wires, and a device. When all are connected, the electrons flow from one electrode through the wire to the device and back to the other electrode.

Current is the number of electrons per second that flow past a point in a circuit and is measured with a device called an ammeter. Current is measured in amperes. An ampere is one coulomb per second.

The **electromotive force** of the battery is the difference in the concentration of electrons on the two electrodes. It is measured in units called volts. A volt is one joule per coulomb.

As electrons flow through a wire, they lose potential energy which is changed into heat energy because of **resistance**. Resistance is the ability of the material to oppose the flow of electrons through it. All substances have some resistance, even if they are a good conductor such as copper. This resistance is measured in units called **ohms**. A thin wire will have more resistance than a thick one because there are less electrons to travel. Resistance also depends upon the length of the wire. The longer the wire, the more resistance it will have.

Potential difference (V), resistance (R), and current (I) form a relationship know as **Ohm's law**: $I = V / R$

A **series circuit** is one where the electrons have only one path along which they can move. When one load in a series circuit goes out, the circuit is open. An example of this is a set of Christmas tree lights that is missing a bulb. None of the bulbs will work.

A **parallel circuit** is one where the electrons have more than one path to move along. If a load goes out in a parallel circuit, the other load will still work because the electrons can still find a way to continue moving along the path.

Competency 13.0   Understand the characteristics and life processes of living organisms

The organization of living systems builds by levels from small to increasingly more large and complex. **Organelles** make up **cells,** which make up **tissues,** which make up **organs**. Groups of organs make up **organ systems**. Organ systems work together to provide life for the **organism.**

There are several ways to describe the difference between living organisms and nonliving entitites:

    1.   Living things are made of cells; they grow, are capable of reproduction and respond to stimuli.
    2.   Living things must adapt to environmental changes or perish.
    3.   Living things carry on metabolic processes. They use energy and make new materials.

The structure of the cell is often related to the cell's function. Root hair cells differ from flower stamens or leaf epidermal cells. They all have different functions.

A single celled organism is called a **protist**. Animal-like protists are called **protozoans.** They do not have chloroplasts. They are usually classified by the way they move for food. Amoebas engulf other protists by flowing around and over them. The paramecium has a hair-like structure that allows it to move back and forth like tiny oars searching for food. The euglena is an example of a protozoan that moves with a tail-like structure called a flagellum.

Plant-like protists have cell walls and float in the ocean. **Bacteria** are the simplest protists. A bacterial cell is surrounded by a cell wall but there is no nucleus inside the cell. Most bacteria do not contain chlorophyll so they do not make their own food. The classification of bacteria is by shape. Cocci are round, bacilli are rod-shaped, and spirilla are spiral-shaped.

Prokaryotic cells consist only of bacteria and blue-green algae. Bacteria were most likely the first cells and date back in the fossil record to 3.5 billion years ago. The important things that put these cells in their own group are

    1. They have no defined nucleus or nuclear membrane. The DNA and ribosomes float freely within the cell.

    2. They have a thick cell wall. This is for protection, to give shape, and to keep the cell from bursting.

3. The cell walls contain amino sugars (glycoproteins). Penicillin works by disrupting the cell wall, which is bad for the bacteria but will not harm the host.

4. Some have a capsule made of polysaccharides which make the bacteria sticky.

5. Some have pili, which is a protein strand. This also allows for attachment of the bacteria and may be used for sexual reproduction (conjugation).

6. Some have flagella for movement.

Eukaryotic cells are found in protists, fungi, plants, and animals. Some features of eukaryotic cells include:

1. They have a nucleus and chromosomes located in the nucleus.

2. They are usually larger than prokaryotic cells.

3. They contain many organelles, which are membrane-bound areas for specific cell functions.

4. They contain a cytoskeleton which provides a protein framework for the cell.

5. They contain cytoplasm to support the organelles and contain the ions and molecules necessary for cell function.

The following are parts of eukaryotic cells:

1. Nucleus - the brain of the cell. The nucleus contains:

chromosomes- DNA, RNA and proteins tightly coiled to conserve space while providing a large surface area.
chromatin - loose structure of chromosomes. Chromosomes are called chromatin when the cell is not dividing.
nucleoli - where ribosomes are made. These are seen as dark spots in the nucleus.
nuclear membrane - contains pores which let RNA out of the nucleus. The nuclear membrane is continuous with the endoplasmic reticulum which allows the membrane to expand or shrink if needed.

2. **Ribosomes** - the site of protein synthesis. Ribosomes may be free floating in the cytoplasm or attached to the endoplasmic reticulum. There may be up to a half a million ribosomes in a cell, depending on how much protein is made by the cell.

3. **Endoplasmic reticulum** - These are folded and provide a large surface area. They are the "roadway" of the cell and allow for the transport of materials. The lumen of the endoplasmic reticulum helps to keep materials out of the cytoplasm and headed in the right direction. The endoplasmic reticulum is capable of building new membrane material. There are two types:

> **Smooth endoplasmic reticulum** - contain no ribosomes on their surface.

> **Rough endoplasmic reticulum** - contain ribosomes on their surface. This form of ER is abundant in cells that make many proteins, like in the pancreas, which produces many digestive enzymes.

4. **Golgi complex or Golgi apparatus** - This structure is stacked to increase surface area. The Golgi Complex functions to sort, modify, and package molecules that are made in other parts of the cell. These molecules are either sent out of the cell or to other organelles within the cell.

5. **Lysosomes** - found mainly in animal cells. These contain digestive enzymes that break down food, substances not needed, viruses, damaged cell components, and eventually the cell itself. It is believed that lysosomes are responsible for the aging process.

6. **Mitochondria** - large organelles that make ATP to supply energy to the cell. Muscle cells have many mitochondria because they use a great deal of energy. The folds inside the mitochondria are called cristae. They provide a large surface where the reactions of cellular respiration occur. Mitochondria have their own DNA and are capable of reproducing themselves if a greater demand is made for additional energy. Mitochondria are found only in animal cells.

7. **Plastids** - found in photosynthetic organisms only. They are similar to the mitochondria due to their double membrane structure. They also have their own DNA and can reproduce if increased capture of sunlight becomes necessary. There are several types of plastids:

> **Chloroplasts** - green; function in photosynthesis. They are capable of trapping sunlight.
> **Chromoplasts** - make and store yellow and orange pigments; they provide color to leaves, flowers and fruits.
> **Amyloplasts** - store starch and are used as a food reserve. They are abundant in roots like potatoes.

8. Cell Wall - Found in plant cells only, it is composed of cellulose and fibers. It is thick enough for support and protection, yet porous enough to allow water and dissolved substances to enter. Cell walls are cemented to each other.

9. Vacuoles - hold stored food and pigments. Vacuoles are very large in plants. This is allows them to fill with water in order to provide turgor pressure. Lack of turgor pressure causes a plant to wilt.

10. Cytoskeleton - composed of protein filaments attached to the plasma membrane and organelles. They provide a framework for the cell and aid in cell movement. They constantly change shape and move about. Three types of fibers make up the cytoskeleton:

> **Microtubules** - largest of the three; make up cilia and flagella for locomotion. Flagella grow from a basal body. Some examples are sperm cells and tracheal cilia. Centrioles are also composed of microtubules. They form the spindle fibers that pull the cell apart into two cells during cell division. Centrioles are not found in the cells of higher plants.

> **Intermediate Filaments** - they are smaller than microtubules but larger than microfilaments. They help the cell to keep its shape.

> **Microfilaments** - smallest of the three, they are made of actin and small amounts of myosin (like in muscle cells). They function in cell movement such as cytoplasmic streaming, endocytosis, and ameboid movement. Microfilaments pinch the two cells apart after cell division, forming two cells.

The purpose of cell division is to provide growth and repair in body (somatic) cells and to replenish or create sex cells for reproduction. There are two forms of cell division. **Mitosis** is the division of somatic cells and **meiosis** is the division of sex cells (eggs and sperm). The table below summarizes the major differences between the two processes.

MITOSIS	MEIOSIS
1. Division of somatic cell	1. Division of sex cells
2. Two cells result from each division	2. Four cells or polar bodies result from each division
3. Chromosome number is identical to parent cells	3. Chromosome number is half the number of parent cells
4. For cell growth and repair	4. Recombinations provide genetic diversity

The following are some related terms to know:

gamete - sex cell or germ cell; eggs and sperm.
chromatin - loose chromosomes; this state is found when the cell is not dividing.
chromosome - tightly coiled, visible chromatin; this state is found when the cell is dividing.
homologues - chromosomes that contain the same information. They are of the same length and contain the same genes.
diploid - 2n number; diploid chromosomes are a pair of chromosomes (somatic cells).
haploid - 1n number; haploid chromosomes are a half of a pair (sex cells).

In mitosis, the cell cycle is the life cycle of the cell. It is divided into two stages: **interphase** and the **mitotic division** where the cell is actively dividing. Interphase is divided into three steps: G1 (growth) period, where the cell is growing and metabolizing, S period (synthesis) where new DNA and enzymes are being made, and the G2 phase (growth) where new proteins and organelles are being made to prepare for cell division. The mitotic stage consists of the stages of mitosis and the division of the cytoplasm.

The stages of mitosis and their events are as follows. Be sure to know the correct order of steps (IPMAT):

1. Interphase - chromatin is loose, chromosomes are replicated, cell metabolism is occurring. Interphase is technically <u>not</u> a stage of mitosis.

2. Prophase - once the cell enters prophase, it proceeds through the following steps continuously with no stopping. The chromatin condenses to become visible chromosomes. The nucleolus disappears and the nuclear membrane (envelope) breaks apart. Mitotic spindles form which will eventually pull the chromosomes apart. They are composed of microtubules. The cytoskeleton breaks down and the spindles are pushed to the poles or opposite ends of the cell by the action of centrioles.

3. Metaphase - kinetechore fibers attach to the chromosomes which causes the chromosomes to line up in the center of the cell (think **m**iddle for **m**etaphase).

4. Anaphase - centromeres split in half and homologous chromosomes separate. The chromosomes are pulled to the poles of the cell, with identical sets at either end.

5. Telophase - two nuclei form with a full set of DNA that is identical to the parent cell. The nucleoli become visible and the nuclear membrane reassembles. A cell plate is visible in plant cells, whereas a cleavage furrow is formed in animal cells. The cell is pinched into two cells. Cytokinesis, or division, of the cytoplasm and organelles occurs.

Meiosis contains the same five stages as mitosis, but is repeated in order to reduce the chromosome number by one half. This way, when the sperm and egg join during fertilization, the haploid number is reached. The steps of meiosis are as follows:

Major function of **meiosis I** - chromosomes are replicated; cells remain diploid:

Prophase I - replicated chromosomes condense and pair with homologues. This forms a tetrad. Crossing over (the exchange of genetic material between homologues to further increase diversity) occurs during Prophase I.
Metaphase I - homologous sets attach to spindle fibers after lining up in the middle of the cell.
Anaphase I - sister chromatids remain joined and move to the poles of the cell.
Telophase I - two new cells are formed; chromosome number is still diploid

Major function of **meiosis II** - to reduce the chromosome number in half:

Prophase II - chromosomes condense.
Metaphase II - spindle fibers form again, sister chromatids line up in center of cell, centromeres divide and sister chromatids separate.
Anaphase II - separated chromosomes move to opposite ends of the cell.
Telophase II - four haploid cells form for each original sperm germ cell. One viable egg cell gets all the genetic information and three polar bodies form with no DNA. The nuclear membrane reforms and cytokinesis occurs.

Photosynthesis is the process by which plants make carbohydrates from the energy of the sun, carbon dioxide, and water. Oxygen is a waste product. Photosynthesis occurs in the chloroplast where the pigment chlorophyll traps sun energy. It is divided into two major steps:

> **Light Reactions** - Sunlight is trapped, water is split, and oxygen is given off. ATP is made and hydrogens reduce NADP to $NADPH_2$. The light reactions occur in light. The products of the light reactions enter into the dark reactions (Calvin cycle).
> **Dark Reactions** - Carbon dioxide enters during the dark reactions which can occur with or without the presence of light. The energy transferred from $NADPH_2$ and ATP allow for the fixation of carbon into glucose.

During times of decreased light, plants break down the products of photosynthesis through cellular **respiration**. Glucose, with the help of oxygen, breaks down and produces carbon dioxide and water as waste. Approximately fifty percent of the products of photosynthesis are used by the plant for energy.

Water travels up the xylem of the plant through the process of **transpiration.** Water sticks to itself (cohesion) and to the walls of the xylem (adhesion). As it evaporates through the stomata of the leaves, the water is pulled up the column from the roots. Environmental factors such as heat and wind increase the rate of transpiration. High humidity will decrease the rate of transpiration.

Competency 14.0   Understand principles related to the inheritance of characteristics

Gregor Mendel is recognized as the father of genetics. His work in the late 1800s is the basis of our knowledge of genetics. Although unaware of the presence of DNA or genes, Mendel realized there were factors (now known as genes) that were transferred from parents to their offspring. Mendel worked with pea plants and fertilized the plants himself, keeping track of subsequent generations which led to the Mendelian laws of genetics. Mendel found that two 'factors' governed each trait, one from each parent. Traits or characteristics came in several forms, known as alleles. For example, the trait of flower color had white alleles and purple alleles. Mendel formed three laws:

Law of dominance - in a pair of alleles, one trait may cover up the allele of the other trait. Example: brown eyes are dominant to blue eyes.

Law of segregation - only one of the two possible alleles from each parent is passed on to the offspring from each parent. (During meiosis, the haploid number insures that half the sex cells get one allele, half get the other).

Law of independent assortment - alleles sort independently of each other. (Many combinations are possible depending on which sperm ends up with which egg. Compare this to the many combinations of hands possible when dealing a deck of cards).

monohybrid cross - a cross.

dihybrid cross - a cross using two traits. More combinations are possible.

Punnet squares are used to show the possible ways that genes combine and indicate probability of the occurrence of a certain genotype or phenotype. One parent's genes are put at the top of the box and the other parent at the side of the box. Genes combine on the square just like numbers that are added in addition tables we learned in elementary school.

A monohybrid cross uses only one trait. There are four possible gene combinations:

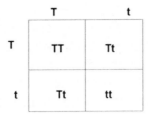

A dihybrid cross uses two traits. There are sixteen possible gene combinations:

R = roundness
r = wrinkled

Y = yellow
y = green

	RY	Ry	rY	ry
RY	RRYY	RRYy	RrYY	RrYy
Ry	RRYy	RRyy	RrYy	Rryy
rY	RrYY	RrYy	rrYY	rrYy
ry	RrYy	Rryy	rrYy	rryy

Since it's not a perfect world, mistakes happen. Inheritable changes in DNA are called **mutations**. Mutations may be errors in replication or a spontaneous rearrangement of one or more segments by factors like radioactivity, drugs, or chemicals. The amount of the change is not as critical as where the change is. Mutations may occur in somatic or sex cells. Usually the ones on sex cells are more dangerous since they contain the basis of all information for the developing offspring. Mutations are not always bad. They are the basis of evolution and, if they make a more favorable variation that enhances the organism's survival, then they are beneficial. But, mutations may also lead to abnormalities, birth defects, and even death. There are several types of mutations; let's suppose a normal sequence was as follows:

Normal -   A B C D E F

Duplication - one gene is repeated.   A B C C D E F

Inversion - a segment of the sequence is flipped around.   A E D C B F

Deletion - a gene is left out.   A B C E F

Insertion or translocation - a segment from another place on the DNA is inserted in the wrong place.   A B C R S D E F

Breakage - a piece is lost.   A B C (DEF is lost)

Nondisjunction – This occurs during meiosis when chromosomes fail to separate properly. One sex cell may get both genes and another may get none. Depending on the chromosomes involved, this may or may not be serious. Offspring end up with either an extra chromosome or are missing one. An example of nondisjunction is Down syndrome, where three of chromosome #21 are present.

The following are some definitions to know:

Dominant - the stronger of the two traits. If a dominant gene is present, it will be expressed- shown by a capital letter.

Recessive - the weaker of the two traits. In order for the recessive gene to be expressed, there must be two recessive genes present.-shown by a lower case letter.

Homozygous - (purebred) having two of the same genes present; an organism may be homozygous dominant with two dominant genes or homozygous recessive with two recessive genes.

Heterozygous - (hybrid) having one dominant gene and one recessive gene. The dominant gene will be expressed due to the law of dominance.

Genotype - the genes the organism has. Genes are represented with letters. AA, Bb, and tt are examples of genotypes.

Phenotype - how the trait is expressed in an organism. Blue eyes, brown hair, and red flowers are examples of phenotypes.

Incomplete dominance - neither gene masks the other; a new phenotype is formed. For example, red flowers and white flowers may have equal strength. A heterozygote (Rr) would have pink flowers. If a problem occurs with a third phenotype, incomplete dominance is occurring.

Codominance - genes may form new phenotypes. The ABO blood grouping is an example of co-dominance. A and B are of equal strength and O is recessive. Therefore, type A blood may have the genotypes of AA or AO, type B blood may have the genotypes of BB or BO, type AB blood has the genotype A and B, and type O blood has two recessive O genes.

Linkage - genes that are found on the same chromosome usually appear together unless crossing over has occurred in meiosis. (Example - blue eyes and blonde hair)

Lethal alleles - these are usually recessive due to the early death of the offspring. If a 2:1 ratio of alleles is found in offspring, a lethal gene combination is usually the reason. Some examples of lethal alleles include sickle cell anemia, Tay-Sachs disease, and cystic fibrosis. Usually the coding for an important protein is affected.

Inborn errors of metabolism - these occur when the protein affected is an enzyme. Examples include PKU (phenylketonuria) and albanism.

Polygenic characters - many alleles code for a phenotype. There may be as many as twenty genes that code for skin color. This is why there is such a variety of skin tones. Another example is height. A couple of medium height may have very tall offspring.

Sex linked traits - the Y chromosome found only in males (XY) carries very little genetic information, whereas the X chromosome found in females (XX) carries very important information. Since men have no second X chromosome to cover up a recessive gene, the recessive trait is expressed more often in men. Women need the recessive gene on both X chromosomes to show the trait. Examples of sex linked traits include hemophilia and color-blindness.

Sex influenced traits - traits are influenced by the sex hormones. Male pattern baldness is an example of a sex influenced trait. Testosterone influences the expression of the gene. Usually men loose their hair due to this trait.

The modern definition of a gene is a unit of genetic information. DNA makes up genes which in turn, make up the chromosomes. DNA is wound tightly around proteins in order to conserve space. The DNA/protein combination makes up the chromosome. DNA controls the synthesis of proteins, thereby controlling the total cell activity. DNA is capable of making copies of itself.

Review of DNA structure:

1. Made of nucleotides; a five carbon sugar, phosphate group and nitrogen base (either adenine, guanine, cytosine or thymine).

2. Consists of a sugar/phosphate backbone which is covalently bonded. The bases are joined down the center of the molecule and are attached by hydrogen bonds which are easily broken during replication.

3. The amount of adenine equals the amount of thymine and the amount of cytosine equals the amount of guanine.

4. The shape is that of a twisted ladder called a double helix. The sugar/phosphates make up the sides of the ladder and the base pairs make up the rungs of the ladder.

Enzymes control each step of the **replication of DNA**. The molecule untwists. The hydrogen bonds between the bases break and serve as a pattern for replication. Free nucleotides found inside the nucleus join to form a new strand. Two new pieces of DNA are formed which are identical. This is a very accurate process. There is only one mistake for every billion nucleotides added. This is because there are enzymes (polymerases) present that proofread the molecule. In eukaryotes, replication occurs in many places along the DNA at once. The molecule may open up at many places like a broken zipper. In prokaryotic circular plasmids, replication begins at a point on the plasmid and goes in both directions until it meets itself.

Base pairing rules are important in determining a new strand of DNA sequence. For example, say our original strand of DNA had the sequence as follows:

1. A T C G G C A A T A G C   This may be called our sense strand as it contains a sequence that makes sense or codes for something. The complementary strand (or other side of the ladder) would follow base pairing rules (A bonds with T and C bonds with G) and would read:

2. T A G C C G T T A T C G   When the molecule opens up and nucleotides add on, the base pairing rules create two new identical strands of DNA

1. A T C G G C A A T A G C   and   2. A T C G G C A A T A G C
   T A G C C G T T A T C G   T A G C C G T T A T C G

It is necessary for cells to manufacture new proteins for growth and repair of the organism. **Protein synthesis** is the process that allows the DNA code to be read and carried out of the nucleus into the cytoplasm in the form of RNA. This is where the ribosomes are found, which are the sites of protein synthesis. The protein is then assembled according to the instructions on the DNA. There are several types of RNA. Familiarize yourself with where they are found and their function.

Messenger RNA - (mRNA) copies the code from DNA in the nucleus and takes it to the ribosomes in the cytoplasm.

Transfer RNA - (tRNA) is free floating in the cytoplasm. Its job is to carry and position amino acids for assembly on the ribosome.

Ribosomal RNA - (rRNA) found in the ribosomes. They make a place for the proteins to be made. rRNA is believed to have many important functions; so much research is currently being done currently in this area.

Along with enzymes and amino acids, the RNA's function is to assist in the building of proteins. There are two stages of protein synthesis:

Transcription - this phase allows for the assembly of mRNA and occurs in the nucleus where the DNA is found. The DNA splits open and the RNA polymerase reads the DNA sequence and "transcribes" a sequence of single stranded mRNA. For example, if the code on the DNA is T A C C T C G T A C G A, the mRNA will make a complementary strand reading: A U G G A G C A U G C U (Remember that uracil replaces thymine in RNA.) The mRNA leaves the nucleus and enters the cytoplasm.

Translation – In the cytoplasm, the mRNA binds to the ribosome. The ribosome is made of rRNA (ribosomal RNA) and protein. The ribosome "reads" three nucleotides of the mRNA at a time. Groups of three mRNA nucleotides are called codons. The codon will eventually code for a specific amino acid to be carried to the ribosome. "Start" codons begin the building of the protein and "stop" codons end translation. When the stop codon is reached, the ribosome stops reading the mRNA and releases the protein and mRNA. The nucleotide sequence is translated to choose the correct amino acid sequence. As the rRNA translates the code at the ribosome, tRNAs which contain an **anticodon** seek out the correct amino acid and bring it back to the ribosome. For example, using the codon sequence from the example above:

the mRNA reads   A U G / G A G / C A U / G C U
the anticodons are U A C / C U C / G U A / C G A
the amino acid sequence would be: Methionine (start) - Glu - His - Ala.

This whole process is accomplished through the assistance of **activating enzymes**. Each of the twenty amino acids has their own enzyme. The enzyme binds the amino acid to the tRNA. When the amino acids get close to each other on the ribosome, they bond together using peptide bonds. The start and stop codons are called nonsense codons. There is one start codon (AUG) and three stop codons. (UAA, UGA and UAG). Addition mutations will cause the whole code to shift, thereby producing the wrong protein or, at times, no protein at all.

Competency 15.0 Understand principles and theories related to biological evolution

Darwin defined the theory of natural selection in the mid-1800s. Through the study of finches on the Galapagos Islands, Darwin theorized that nature selects the traits that are advantageous to the organism. Those that do not possess the desirable trait die and do not pass on their genes. Those more fit to survive reproduce, thus increasing that gene in the population. Darwin listed four principles to define natural selection:

1. The individuals in a certain species vary from generation to generation.
2. Some of the variations are determined by the genetic makeup of the species.
3. More individuals are produced than will survive.
4. Some genes allow for better survival of an animal.

Causes of evolution - Certain factors increase the chances of variability in a population, thus leading to evolution. Items that increase variability include mutations, sexual reproduction, immigration, and large population. Items that decrease variation would be natural selection, emigration, small population, and random mating.

Sexual selection - Genes that happen to come together determine the makeup of the gene pool. Animals that use mating behaviors may be successful or unsuccessful. An animal that lacks attractive plumage or has a weak mating call will not attract the female, thereby eventually limiting that gene in the gene pool. Mechanical isolation, where sex organs do not fit the female, has an obvious disadvantage.

Competency 16.0   Understand characteristics of populations, communities, ecosystems, and biomes

Ecology is the study of organisms, where they live and their interactions with the environment. A **population** is a group of the same species in a specific area. A **community** is a group of populations residing in the same area. Communities that are ecologically similar with regard to temperature, rainfall, and the species that live there are called **biomes**. Specific biomes include:

Marine - covers 75% of the earth. This biome is organized by the depth of the water. The intertidal zone is from the tide line to the edge of the water. The littoral zone is from the water's edge to the open sea. It includes coral reef habitats and is the most densely populated area of the marine biome. The open sea zone is divided into the epipelagic zone and the pelagic zone. The epipelagic zone receives more sunlight and has a larger number of species. The ocean floor is called the benthic zone and is populated with bottom feeders.

Tropical Rain Forest - temperature is constant (25°C), rainfall exceeds 200 cm per year. Located around the area of the equator, the rain forest has abundant, diverse species of plants and animals.

Savanna - temperatures range from 0–25°C, depending on the location. Rainfall is from 90 to 150 cm per year. Plants include shrubs and grasses. The savanna is a transitional biome between the rain forest and the desert.

Desert - temperatures range from 10–38°C degrees. Rainfall is under 25 cm per year. Plant species include xerophytes and succulents. Lizards, snakes and small mammals are common animals.

Temperate Deciduous Forest - temperature ranges from –24 to 38°C. Rainfall is between 65 to 150 cm per year. Deciduous trees are common, as well as deer, bear and squirrels.

Taiga - temperatures range from –24 to 22°C. Rainfall is between 35 to 40 cm per year. Taiga is located very north and very south of the equator, getting close to the poles. Plant life includes conifers and plants that can withstand harsh winters. Animals include weasels, mink, and moose.

Tundra - temperatures range from –28 to 15°C. Rainfall is limited, ranging from 10 to 15 cm per year. The tundra is located even further north and south than the taiga. Common plants include lichens and mosses. Animals include polar bears and musk ox.

Polar or Permafrost - temperature ranges from –40 to 0°C. It rarely gets above freezing. Rainfall is below 10 cm per year. Most water is bound up as ice. Life is limited.

Succession is an orderly process of replacing a community that has been damaged or beginning one where no life previously existed. Primary succession occurs after a community has been totally wiped out by a natural disaster or where life never existed before, as in a flooded area. Secondary succession takes place in communities that were once flourishing but were disturbed by some source, either man or nature, but were not totally stripped. A climax community is a community that is established and flourishing.

There are a number of **feeding relationships** between different species:

Parasitism - two species that occupy a similar place; the parasite benefits from the relationship, the host is harmed.

Commensalism - two species that occupy a similar place; neither species is harmed or benefits from the relationship.

Mutualism (symbiosis) - two species that occupy a similar place; both species benefit from the relationship.

Competition - two species that occupy the same habitat or eat the same food are said to be in competition with each other.

Predation - animals that eat other animals are called predators. The animals they feed on are called the prey. Population growth depends upon competition for food, water, shelter, and space. The amount of predators determines the amount of prey which, in turn, affects the number of predators.

Carrying Capacity - this is the total amount of life a habitat can support. Once the habitat runs out of food, water, shelter, or space, the carrying capacity decreases and then stabilizes.

There are **ecological problems.** Nonrenewable resources are fragile and must be conserved for use in the future. Man's impact and knowledge of conservation will control our future.

Biological magnification refers to the fact that chemicals and pesticides accumulate along the food chain. Tertiary consumers have more accumulated toxins than animals at the bottom of the food chain.

Three major crops feed the world (rice, corn, wheat). The planting of these foods wipe out habitats and push animals residing there into other habitats causing overpopulation or extinction. This is called **simplification of the food web.**

Strip mining and the overuse of oil reserves have depleted **fuel sources**. At the current rate of consumption, conservation or alternate fuel sources will guarantee our future fuel sources.

Although technology gives us many advances, **pollution** is a side effect of production. Waste disposal and the burning of fossil fuels have polluted our land, water and air. Global warming and acid rain are two results of the burning of hydrocarbons and sulfur.

Rainforest depletion and the use of fossil fuels and aerosols have caused an increase in carbon dioxide production. This leads to a decrease in the amount of oxygen which is directly proportional to the amount of ozone. As the ozone layer depletes, more heat enters our atmosphere and is trapped. This causes an overall warming effect which may eventually melt polar ice caps, causing a rise in water levels and changes in climate which will affect weather systems world-wide. This is referred to as **global warming**.

Construction of homes to house people in our overpopulated world has caused the destruction of habitat for other animals, leading to their extinction. These animals are considered **endangered species.**

Biotic factors are living things in an ecosystem: plants, animals, bacteria, fungi, etc. If one population in a community increases, it affects the ability of another population to succeed by limiting the available amount of food, water, shelter and space.

Abiotic factors are nonliving aspects of an ecosystem: soil quality, rainfall, and temperature. Changes in climate and soil can cause effects at the beginning of the food chain, thus limiting or accelerating the growth of populations.

SUBAREA V.   **EARTH AND SPACE SCIENCE**

Competency 17.0   Understand geological history and processes related to the changing earth

Modern Geology

The biological history of the earth is partitioned into four major eras that are further divided into major periods. The latter periods are refined into groupings called Epochs.

Earth's history extends over more than four billion years and is reckoned in terms of a scale. Paleontologists who study the history of the Earth have divided this huge period of time into four large time units called eons. Eons are divided into smaller units of time called eras. An era refers to a time interval in which particular plants and animals were dominant or present in great abundance. The end of an era is most often characterized by (1) a general uplifting of the crust, (2) the extinction of the dominant plants or animals, and (3) the appearance of new life forms. Each era is divided into several smaller divisions of time called periods. Some periods are divided into smaller time units called epochs.

Uniformitarianism is a fundamental concept in modern geology. It simply states that the physical, chemical, and biological laws that operated in the geologic past operate in the same way today. The forces and processes that we observe presently shaping our planet have been at work for a very long time. This idea is commonly stated as "the present is the key to the past." It replaced the concept of **catastrophism** which says the earth was shaped by catastrophic events of a short term nature.

Geologic Dating

Estimates of the earth's age have been made possible with the discovery of **radioactivity** and the invention of instruments that can measure the amount of radioactivity in rocks. The use of radioactivity to make accurate determinations of earth's age is called absolute dating. This process depends upon comparing the amount of radioactive material in a rock with the amount that has decayed into another element. Studying the radiation given off by atoms of radioactive elements is the most accurate method of measuring the earth's age.

The time required for one half of a given amount of a radioactive element to decay is called the half-life of that element or compound. The half-life of carbon-14 is 5730 years. Geologists commonly use carbon dating to calculate the age of a fossil substance.

The determination of the age of rocks by cataloging their composition has been outmoded since the middle 1800s. Today, a sequential history can be determined by the fossil content (principle of fossil succession) of a rock system as well as its superposition within a range of systems. This classification process was termed stratigraphy and permitted the construction of a geologic column in which rock systems are arranged in their correct chronological order.

Minerals

Minerals are natural, nonliving solids with a definite chemical composition and a crystalline structure. **Ores** are minerals or rock deposits that can be mined for a profit. **Rocks** are earth materials made of one or more minerals. A **rock facies** is a rock group that differs from comparable rocks (as in composition, age or fossil content).

There are over 3000 minerals in Earth's crust. Minerals are classified by composition. The major groups of minerals are silicates, carbonates, oxides, sulfides, sulfates, and halides. The largest group of minerals is the silicates. Silicates are made of silicon, oxygen, and one or more other elements.

Igneous Rocks

The first type of rock formed is **igneous rock**. It forms when molten material called magma cools. When molten rock pours out onto the surface of Earth, it is called lava.

Intrusive rock includes any igneous rock that was formed below the earth's surface. Batholiths are the largest structures of intrusive type rock and are composed of near granite materials; they are the core of the Sierra Nevada Mountains.

Extrusive rock includes any igneous rock that was formed at the earth's surface. **Dikes** are old lava tubes formed when magma entered a vertical fracture and hardened. Sometimes magma squeezes between two rock layers and hardens into a thin horizontal sheet called a **sill**. A **laccolith** is formed in much the same way as a sill, but the magma that creates a laccolith is very thick and does not flow easily. It pools and forces the overlying strata creating an obvious surface dome.

Weathering and Soil

Soils are composed of particles of sand, clay, various minerals, tiny living organisms, and humus, plus the decayed remains of plants and animals. Soils are divided into three classes according to their texture. These classes are sandy soils, clay soils, and loamy soils.

Sandy soils are gritty, and their particles do not bind together firmly. Sandy soils are porous—water passes through them rapidly. Sandy soils do not hold much water.

Clay soils are smooth and greasy, their particles bind together firmly. Clay soils are moist and usually do not allow water to pass through easily.

Loamy soils feel somewhat like velvet and their particles clump together. Loamy soils are made up of sand, clay, and silt. Loamy soils holds water but some water can pass through.

In addition to three main classes, soils are further grouped into three major types based upon their composition. These groups are pedalfers, pedocals, and laterites.

Pedalfers form in the humid, temperate climate of the eastern United States. Pedalfer soils contain large amounts of iron oxide and aluminum-rich clays, making the soil a brown to reddish brown color. This soil supports forest type vegetation.

Pedocals are found in the western United States where the climate is dry and temperate. These soils are rich in calcium carbonate. This type of soil supports grasslands and brush vegetation.

Laterites are found where the climate is wet and tropical. Large amounts of water flows through this soil. Laterites are red-orange soils rich in iron and aluminum oxides. There is little humus and this soil is not very fertile.

Erosion is the inclusion and transportation of surface materials by another moveable material, usually water, wind, or ice. The most important cause of erosion is running water. Streams, rivers, and tides are constantly at work removing weathered fragments of bedrock and carrying them away from their original location.

A stream erodes bedrock by the grinding action of the sand, pebbles and other rock fragments. This grinding against each other is called abrasion. Streams also erode rocks by dissolving or absorbing their minerals. Limestone and marble are readily dissolved by streams.

The breaking down of rocks at or near to the earth's surface is known as **weathering**. Weathering breaks down these rocks into smaller and smaller pieces. There are two types of weathering: physical weathering and chemical weathering.

Physical weathering is the process by which rocks are broken down into smaller fragments without undergoing any change in chemical composition. Physical weathering is mainly caused by freezing water, the expansion of rock, and the activities of plants and animals.

Frost wedging is the cycle of daytime thawing and refreezing at night. This cycle causes large rock masses, especially the rocks exposed on mountain tops, to be broken into smaller pieces.

The peeling away of the outer layers from a rock is called exfoliation. Rounded mountain tops are called exfoliation domes and have been formed in this way.

Chemical weathering is the breaking down of rocks through changes in their chemical composition. An example would be the change of feldspar in granite to clay. Water, oxygen, and carbon dioxide are the main agents of chemical weathering. When water and carbon dioxide combine chemically, they produce a weak acid that breaks down rocks.

Sedimentary Rocks

When fluid sediments are transformed into solid sedimentary rocks, the process is known as **lithification**. One very common process affecting sediments is compaction where the weights of overlying materials compress and compact the deeper sediments. The compaction process leads to cementation. **Cementation** is when sediments are converted to sedimentary rock.

Metamorphic Rocks

Metamorphic rocks are formed by high temperatures and great pressures. The process by which the rocks undergo these changes is called metamorphism. The outcome of metamorphic changes include deformation by extreme heat and pressure, compaction, destruction of the original characteristics of the parent rock, bending and folding while in a plastic stage, and the emergence of completely new and different minerals due to chemical reactions with heated water and dissolved minerals.

Metamorphic rocks are classified into two groups, foliated (leaf-like) rocks and unfoliated rocks. Foliated rocks consist of compressed, parallel bands of minerals, which give the rocks a striped appearance. Examples of such rocks include slate, schist, and gneiss. Unfoliated rocks are not banded and examples of such include quartzite, marble, and anthracite rocks.

Volcanic Activity

Volcanism is the term given to the movement of magma through the crust and its emergence as lava onto the earth's surface. Volcanic mountains are built up by successive deposits of volcanic materials.

An active volcano is one that is presently erupting or building to an eruption. A dormant volcano is one that is between eruptions but still shows signs of internal activity that might lead to an eruption in the future. An extinct volcano is said to be no longer capable of erupting. Most of the world's active volcanoes are found along the rim of the Pacific Ocean, which is also a major earthquake zone. This curving belt of active faults and volcanoes is often called the *Ring of Fire*.

The world's best known volcanic mountains include: Mount Etna in Italy and Mount Kilimanjaro in Africa. The Hawaiian Islands are actually the tops of a chain of volcanic mountains that rise from the ocean floor.

There are three types of volcanic mountains: shield volcanoes, cinder cones and composite volcanoes.

Shield volcanoes are associated with quiet eruptions. Lava emerges from the vent or opening in the crater and flows freely out over the earth's surface until it cools and hardens into a layer of igneous rock. A repeated lava flow builds this type of volcano into the largest volcanic mountain. Mauna Loa, found in Hawaii, is the largest volcano on earth.

Cinder cone volcanoes are associated with explosive eruptions as lava is hurled high into the air in a spray of droplets of various sizes. These droplets cool and harden into cinders and particles of ash before falling to the ground. The ash and cinder pile up around the vent to form a steep, cone-shaped hill called the cinder cone. Cinder cone volcanoes are relatively small but may form quite rapidly.

Composite volcanoes are described as being built by both lava flows and layers of ash and cinders. Mount Fuji in Japan, Mount St. Helens in Washington, USA and Mount Vesuvius in Italy are all famous composite volcanoes.

A **caldera** is normally formed by the collapse of the top of a volcano. This collapse can be caused by a massive explosion that destroys the cone and empties most if not all of the magma chamber below the volcano. The cone collapses into the empty magma chamber forming a caldera.

An inactive volcano may have magma solidified in its pipe. This structure, called a volcanic neck, is resistant to erosion and today may be the only visible evidence of the past presence of an active volcano.

Plate Tectonics

Data obtained from many sources led scientists to develop the theory of plate tectonics. This theory is the most current model that explains not only the movement of the continents, but also the changes in the earth's crust caused by internal forces.

Plates are rigid blocks of earth's crust and upper mantle. These rigid solid blocks make up the lithosphere. The earth's lithosphere is broken into nine large sections and several small ones. These moving slabs are called plates. The major plates are named after the continents they are "transporting."

The plates float on and move with a layer of hot, plastic-like rock in the upper mantle. Geologists believe that the heat currents circulating within the mantle cause this plastic zone of rock to slowly flow, carrying along the overlying crustal plates.

Movement of these crustal plates creates areas where the plates diverge as well as areas where the plates converge. A major area of divergence is located in the mid-Atlantic. Currents of hot mantle rock rise and separate at this point of divergence creating new oceanic crust at the rate of 2 to 10 centimeters per year. Convergence is when the oceanic crust collides with either another oceanic plate or a continental plate. The oceanic crust sinks, forming an enormous trench and generating volcanic activity. Convergence also includes continent-to-continent plate collisions. When two plates slide past one another a transform fault is created.

These movements produce many major features of the earth's surface, such as mountain ranges, volcanoes, and earthquake zones. Most of these features are located at plate boundaries, where the plates interact by spreading apart, pressing together, or sliding past each other. These movements are very slow, averaging only a few centimeters a year.

Boundaries form between spreading plates where the crust is forced apart in a process called rifting. Rifting generally occurs at mid-ocean ridges. Rifting can also take place within a continent, splitting the continent into smaller landmasses that drift away from each other, thereby forming an ocean basin between them. The Red Sea is a product of rifting. As the seafloor spreading takes place, new material is added to the inner edges of the separating plates. In this way the plates grow larger, and the ocean basin widens. This is the process that broke up the super-continent Pangaea and created the Atlantic Ocean.

Boundaries between plates that are colliding are zones of intense crustal activity. When a plate of ocean crust collides with a plate of continental crust, the more dense oceanic plate slides under the lighter continental plate and plunges into the mantle. This process is called **subduction**, and the site where it takes place is called a subduction zone. A subduction zone is usually seen on the sea-floor as a deep depression called a trench.

The crustal movement, which is identified by plates sliding sideways past each other, produces a plate boundary characterized by major faults that are capable of unleashing powerful earth-quakes. The San Andreas Fault forms such a boundary between the Pacific Plate and the North American Plate.

Mountain Building (Orogeny)

A mountain is terrain that has been raised high above the surrounding landscape by volcanic action, or some form of tectonic plate collisions. The plate collisions could be intercontinental or ocean floor collisions with a continental crust (subduction). The physical composition of mountains would include igneous, metamorphic, or sedimentary rocks; some may have rock layers that are tilted or distorted by plate collision forces.

There are many different types of mountains. The physical attributes of a mountain range depends upon the angle at which plate movement thrusts layers of rock to the surface. Many mountains (Adirondacks, Southern Rockies) were formed along high angle faults.

Folded mountains (Alps, Himalayas) are produced by the folding of rock layers during their formation. The Himalayas are the highest mountains in the world and contain Mount Everest which rises almost 9 km above sea level. The Himalayas were formed when India collided with Asia. The movement which created this collision is still in process at the rate of a few centimeters per year.

Fault-block mountains (Utah, Arizona, and New Mexico) are created when plate movement produces tension forces instead of compression forces. The area under tension produces normal faults and rock along these faults is displaced upward.

Dome mountains are formed as magma tries to push up through the crust but fails to break the surface. Dome mountains resemble a huge blister on the earth's surface.

Upwarped mountains (Black Hills of S.D.) are created in association with a broad arching of the crust. They can also be formed by rock thrusting upward along high angle faults.

Crustal Deformation

Faults are fractures in the earth's crust which have been created by either tension or compressive forces transmitted through the crust. These forces are produced by the movement of separate blocks of crust.

Faultings are categorized on the basis of the relative movement between the blocks on both sides of the fault plane. The movement can be horizontal, vertical or oblique.

A dip-slip fault occurs when the movement of the plates is vertical and opposite. The displacement is in the direction of the inclination, or dip, of the fault. Dip-slip faults are classified as normal faults when the rock above the fault plane moves down relative to the rock below.

Reverse faults are created when the rock above the fault plane moves up relative to the rock below. Reverse faults having a very low angle to the horizontal are also referred to as thrust faults.

Faults in which the dominant displacement is horizontal movement along the trend or strike (length) of the fault are called **strike-slip faults**. When a large strike-slip fault is associated with plate boundaries it is called a **transform fault**. The San Andreas Fault in California is a well-known transform fault.

Faults that have both vertical and horizontal movement are called **oblique-slip faults**.

Glaciation

A continental glacier covered a large part of North America during the most recent ice age. Evidence of this glacial coverage remains as abrasive grooves, large boulders from northern environments dropped in southerly locations, glacial troughs created by the rounding out of steep valleys by glacial scouring, and the remains of glacial sources called **cirques** that were created by frost wedging the rock at the bottom of the glacier. Remains of plants and animals found in warm climate have been discovered in the moraines and outwash plains help to support the theory of periods of warmth during the past ice ages.

The Ice Age began about 2–3 million years ago. This age saw the advancement and retreat of glacial ice over millions of years. Theories relating to the origin of glacial activity include plate tectonics, where it can be demonstrated that some continental masses, now in temperate climates, were at one time blanketed by ice and snow. Another theory involves changes in the earth's orbit around the sun, changes in the angle of the earth's axis, and the wobbling of the earth's axis. Support for the validity of this theory has come from deep ocean research that indicates a correlation between climatic sensitive micro-organisms and the changes in the earth's orbital status.

There are two main types of glaciers: valley glaciers and continental glaciers. Erosion by valley glaciers is characteristic of U-shaped erosion. They produce sharp peaked mountains such as the Matterhorn in Switzerland. Erosion by continental glaciers often rides over mountains in their paths leaving smoothed, rounded mountains and ridges.

Competency 18.0   Understand characteristics and properties of the hydrosphere

The Hydrosphere

Water that falls to earth in the form of rain and snow is called **precipitation.** Precipitation is part of a continuous process in which water at the earth's surface evaporates, condenses into clouds, and returns to earth. This process is termed the **water cycle**. The water located below the surface is called groundwater.

Oceans

Seventy percent of the earth's surface is covered with saltwater. The mass of this saltwater is about 1.4×10^{24} grams. The ocean waters continuously circulate among different parts of the hydrosphere. There are seven major oceans: the North Atlantic Ocean, South Atlantic Ocean, North Pacific Ocean, South Pacific Ocean, Indian Ocean, Arctic Ocean, and the Antarctic Ocean.

Pure water makes up about 96.5% of ocean water. The remaining portion is made up of dissolved solids. The concentration of these dissolved solids determines the water's salinity.

Salinity is the number of grams of these dissolved salts in 1,000 grams of sea water. The average salinity of ocean water is about 3.5%. In other words, one kilogram of sea water contains about 35 grams of salt. Sodium chloride (NaCl) or table salt is the most abundant of the dissolved salts. The dissolved salts also include smaller quantities of magnesium chloride, magnesium and calcium sulfates, and traces of several other salt elements. Salinity varies throughout the world oceans; the total salinity of the oceans varies from place to place and also varies with depth. Salinity is low near river mouths where the ocean mixes with fresh water, and salinity is high in areas of high evaporation rates.

The temperature of the ocean water varies with different latitudes and with ocean depths. Ocean water temperature is about constant to depths of 90 meters. The temperature of surface water will drop rapidly from 28°C at the equator to --2°C at the North Pole and South Pole. The freezing point of sea water is lower than the freezing point of pure water (0°C).

World weather patterns are greatly influenced by ocean surface currents in the upper layer of the ocean. These currents continuously move along the ocean surface in specific directions. Ocean currents that flow deep below the surface are called sub-surface currents. These currents are influenced by such factors as the location of landmasses in the current's path and the earth's rotation.

Surface currents are caused by winds and are classified by temperature. Cold currents originate in the polar regions and flow through surrounding water that is measurably warmer. Those currents with a higher temperature than the surrounding water are called warm currents and can be found near the equator. These currents follow swirling routes around the ocean basins and the equator.

The Gulf Stream and the California Current are the two main surface currents that flow along the coastlines of the United States. The Gulf Stream is a warm current in the Atlantic Ocean that carries warm water from the equator to the northern parts of the Atlantic Ocean. Benjamin Franklin studied and named the Gulf Stream. The California Current is a cold current that originates in the Arctic regions and flows southward along the west coast of the United States.

Differences in water density also create ocean currents. Water found near the bottom of oceans is the coldest and the densest. Water tends to flow from a denser area to a less dense area. Currents that flow because of a difference in the density of the ocean water are called density currents. Water with a higher salinity is denser than water with a lower salinity. Water that has salinity different from the surrounding water may form a density current.

Causes and Effects of Waves

The movement of ocean water is caused by the wind, the sun's heat energy, the earth's rotation, the moon's gravitational pull on earth, and by underwater earthquakes. Most ocean waves are caused by the impact of winds. Wind blowing over the surface of the ocean transfers energy by way of friction to the water and causes waves to form.

Waves are also formed by seismic activity on the ocean floor. A wave formed by an earthquake is called a seismic sea wave. These powerful waves can be very destructive, with wave heights increasing to 30 m or more near the shore. The crest of a wave is its highest point. The trough of a wave is its lowest point. The distance from wave top to wave top is the wavelength. The wave period is the time between the passings of two successive waves.

Shoreline

The shoreline is the boundary where land and sea meet. Shorelines mark the average position of sea level, which is the average height of the sea without consideration of tides and waves. Shorelines are classified according to the way they were formed. The three types of shorelines are: submerged, emergent, and neutral. When the sea has risen, or the land has sunk, a **submerged shoreline** is created. An **emergent shoreline** occurs when sea falls or the land rises. A **neutral shoreline** does not show the features of a submerged or an emergent shoreline. A neutral shoreline is usually observed as a flat and broad beach.

Waves approaching the beach at a slight angle create a current of water that flows parallel to the shore. This longshore current carries loose sediment almost like a river of sand. A spit is formed when a weak longshore current drops its load of sand as it turns into a bay.

Rip currents are narrow currents that flow seaward at a right angle to the shoreline.   These currents are very dangerous to swimmers. Most of the beach sands are composed of grains of resistant material like quartz and orthoclase but coral or basalt are found in some locations. Many beaches have rock fragments that are too large to be classified as sand.

Competency 19.0   Understand the earth's atmosphere, weather, and climate

Air masses moving toward or away from the earth's surface are called air currents. Air moving parallel to earth's surface is called **wind**. Weather conditions are generated by winds and air currents carrying large amounts of heat and moisture from one part of the atmosphere to another. Wind speeds are measured by instruments called anemometers.

The wind belts in each hemisphere consist of convection cells that encircle earth like belts. There are three major wind belts on earth: (1) trade winds, (2) prevailing westerlies, and (3) polar easterlies. Wind belt formation depends on the differences in air pressures that develop in the doldrums, the horse latitudes, and the polar regions. The doldrums surround the equator. Within this belt heated air usually rises straight up into earth's atmosphere. The horse latitudes are regions of high barometric pressure with calm and light winds. The polar regions contain cold dense air that sinks to the Earth's surface.

Winds caused by local temperature changes include **sea breezes** and **land breezes**. Sea breezes are caused by the unequal heating of the land and an adjacent, large body of water. Land heats up faster than water. The movement of cool ocean air toward the land is called a sea breeze. Sea breezes usually begin blowing about mid-morning and end about sunset. A breeze that blows from the land to the ocean or a large lake is called a land breeze.

Monsoons are huge wind systems that cover large geographic areas and that reverse direction seasonally. The monsoons of India and Asia are examples of these seasonal winds. They alternate wet and dry seasons. As denser cooler air over the ocean moves inland, a steady seasonal wind called a summer or wet monsoon is produced.

Clouds are visible masses of water droplets or frozen crystals. The main types of clouds are

Cirrus clouds - white and feathery; high in the sky

Cumulus – thick, white, fluffy

Stratus – layers of clouds cover most of the sky

Nimbus – heavy, dark clouds that represent thunderstorm clouds

The air temperature at which water vapor begins to condense is called the **dew point. Relative humidity** is the actual amount of water vapor in a certain volume of air compared to the maximum amount of water vapor this air could hold at a given temperature.

A **thunderstorm** is a brief, local storm produced by the rapid upward movement of warm, moist air within a cumulo-nimbus cloud. Thunderstorms always produce lightning and thunder, and are accompanied by strong wind gusts and heavy rain or hail.

A severe storm with swirling winds that may reach speeds of hundreds of km per hour is called a **tornado**. Such a storm is also referred to as a "twister." The sky is covered by large cumulo-nimbus clouds and violent thunderstorms; a funnel-shaped swirling cloud may extend downward from a cumulo-nimbus cloud and reach the ground. Tornadoes are storms that leave a narrow path of destruction on the ground.

A swirling, funnel-shaped cloud that **extends** downward and touches a body of water is called a **waterspout.**

Hurricanes are storms that develop when warm, moist air carried by trade winds rotate around a low-pressure "eye." A large, rotating, low-pressure system accompanied by heavy precipitation and strong winds is called a tropical cyclone (better known as a hurricane). In the Pacific region, a hurricane is called a typhoon.

Storms that occur only in the winter are known as blizzards or ice storms. A **blizzard** is a storm with strong winds, blowing snow and frigid temperatures. An **ice storm** consists of falling rain that freezes when it strikes the ground, covering everything with a layer of ice.

Competency 20.0   Understand components of the solar system and universe and their interactions

Origin of Universe and Solar System

Two main hypotheses of the origin of the solar system are the tidal hypothesis and the condensation hypothesis.

The **tidal hypothesis** proposes that the solar system began with a near collision of the sun and a large star. Some astronomers believe that as these two stars passed each other, and the great gravitational pull of the large star extracted hot gases out of the sun. The mass from the hot gases started to orbit the sun, which began to cool then condensing into the nine planets. Few astronomers support this theory.

The condensation hypothesis proposes that the solar system began with rotating clouds of dust and gas. Condensation occurred in the center forming the sun and the smaller parts of the cloud formed the nine planets. This theory is more is widely accepted by astronomers.

Two main theories to explain the origins of the universe are the Big Bang and the Steady State theory.

The **Big Bang** has been widely accepted by many astronomers. It states that the universe originated from a magnificent explosion spreading mass, matter and energy into space. The galaxies formed from this material as it cooled during the next half-billion years.

The Steady State theory has less acceptance. It states that the universe is continuously being renewed. Galaxies move outward and new galaxies replace the older galaxies. Astronomers have not found any evidence to prove this theory.

The future of the universe is predicted with the Oscillating Universe hypothesis. It states that the universe will oscillate or expand and contract. Galaxies will move away from one another and will in time slow down and stop. Then a gradual moving toward each other will again activate the explosion or the Big Bang.

The Sun

The **sun** is the nearest star to earth. It produces solar energy by the process of nuclear fusion, that is, the conversion of hydrogen to helium. Energy flows out of the core of the sun to the surface, then radiates into space.

Parts of the sun include: (1) **core**: the inner portion of the sun where fusion takes place, (2) **photosphere**: considered the surface of the sun which produces **sunspots** (cool, dark areas that can be seen on its surface), (3) **chromosphere**: hydrogen gas causes this portion to be red in color (also found here are solar flares or sudden brightness of the chromosphere) and solar prominences (gases that shoot outward from the chromosphere), and (4) **corona**, the transparent area of sun visible only during a total eclipse.

Solar radiation is energy traveling from the sun that radiates into space. **Solar flares** produce excited protons and electrons that shoot outward from the chromosphere at great speeds reaching earth. These particles disturb radio reception and also affect the magnetic field on earth.

Telescopes

Galileo was the first person to use telescopes to observe the solar system. He invented the first refracting telescope. A **refracting telescope** uses lenses to bend light rays to create the image.

Sir Isaac Newton invented the **reflecting telescope** which uses a concave mirror to gather light rays and produce an image.

The world's largest telescope is located in Mauna Kea, Hawaii. It uses multiple mirrors to gather light rays.

The **Hubble Space telescope** uses a **single-reflector mirror**. It allows astronomers to observe objects seven times farther away than telescopes on earth. Even those objects that are 50 times fainter can be viewed better than by any telescope on Earth. There are future plans to make repairs and install new mirrors and other equipment on the Hubble Space telescope.

Refracting and reflecting telescopes are considered **optical telescopes** since they gather visible light and focus it to produce images. A different type of telescope that collects invisible radio waves created by the sun and stars is called a **radio telescope.**

Radio telescopes consists of a reflector or dish with special receivers. The reflector collects radio waves that are created by the sun and stars. Using a radio telescope has many advantages. They can receive signals 24 hours a day, can operate in any kind of weather, and dust particles or clouds do not interfere with its performance. The most impressive aspect of the radio telescope is its ability to detect objects from great distances in space.

The world's largest radio telescope is located in Arecibo, Puerto Rico. It has a collecting dish antenna of more than 300 meters in diameter.

Spectroscopes

The **spectroscope** is a device or an attachment for telescopes that is used to separate white light into a series of different colors by wave lengths. This series of colors of light is called a **spectrum**. A **spectrograph** can photograph a spectrum. Wavelengths of light have distinctive colors. The color red has the longest wavelength and violet has the shortest wavelength. Wavelengths are arranged to form an **electromagnetic spectrum**. They range from very long radio waves to very short gamma rays. Visible light covers a small portion of the electromagnetic spectrum. Spectroscopes observe the spectra, temperatures, pressures, and also the movement of stars. The spectra of stars indicates if they are moving toward, or away, from earth.

If a star is moving towards earth, light waves compress and the wavelengths of light seem shorter. This will cause the entire spectrum to move towards the blue or violet end of the spectrum.

If a star is moving away from earth, light waves expand and the wavelengths of light seem longer. This will cause the entire spectrum to move towards the red end of the spectrum.

Astronomical Measurements

The three formulas astronomers use for calculating distances in space are the (1) **astronomical unit** (AU), (2) **light year** (ly), and (3) **parsec** (pc). It is important to remember that these formulas are measured in distances not time.

The distance between the earth and the sun is about 150×10^6 km. This distance is known as an astronomical unit or AU. The distance light travels in one year is a light year (9.5×10^{12} km). Large distances are measured in parsecs. One parsec equals 3.26 light years.

Planets

There are eight established planets in our solar system; Mercury, Venus, Earth, Mars, Jupiter, Saturn, Uranus, and Neptune. Pluto was an established planet in our solar system. But as of the summer of 2006, its status is being reconsidered. The planets are divided into two groups based on distance from the sun. The inner planets are Mercury, Venus, Earth, and Mars. The outer planets are Jupiter, Saturn, Uranus, and Neptune.

Mercury is the closest planet to the sun. Its surface has craters and rocks. The atmosphere is composed of hydrogen, helium and sodium. Mercury was named after the Roman messenger god.

Venus has a slow rotation compared to Earth's rotation. Venus and Uranus rotate in opposite directions from the other planets. This opposite rotation is called retrograde rotation. The surface of Venus is not visible due to the extensive cloud cover. The atmosphere is composed mostly of carbon dioxide. Sulfuric acid droplets in the dense cloud cover give Venus a yellow appearance. Venus has a greater greenhouse effect than observed on Earth. The dense clouds combined with carbon dioxide trap heat. Venus was named after the Roman goddess of love.

Earth is considered a water planet with 70% of its surface covered by water. Gravity holds the masses of water in place. The different temperatures observed on earth allow for the different states (solid, liquid, gas) of water to exist. The atmosphere is composed mainly of oxygen and nitrogen. Earth is the only planet that is known to support life.

Mars has a surface that contains numerous craters, active and extinct volcanoes, ridges, and valleys with extremely deep fractures. Iron oxide found in the dusty soil makes the surface seem rust-colored and the skies pink in color. The atmosphere is composed of carbon dioxide, nitrogen, argon, oxygen, and water vapor. Mars has polar regions with ice caps composed of water. Mars has two satellites. Mars was named after the Roman war god.

Jupiter is the largest planet in the solar system. Jupiter has 16 moons. The atmosphere is composed of hydrogen, helium, methane and ammonia. There are white colored bands of clouds indicating rising gas and dark colored bands of clouds indicating descending gases. The gas movement is caused by heat resulting from the energy of Jupiter's core. Jupiter has a Great Red Spot that is thought to be a hurricane-type cloud. Jupiter has a strong magnetic field.

Saturn is the second largest planet in the solar system. Saturn has rings of ice, rock, and dust particles circling it. Saturn's atmosphere is composed of hydrogen, helium, methane, and ammonia. Saturn has 20 plus satellites. Saturn was named after the Roman god of agriculture.

Uranus is a gaseous planet. It has 10 dark rings and 15 satellites. Its atmosphere is composed of hydrogen, helium, and methane. Uranus was named after the Greek god of the heavens.

Neptune is another gaseous planet with an atmosphere consisting of hydrogen, helium, and methane. Neptune has 3 rings and 2 satellites. Neptune was named after the Roman sea god because its atmosphere is the same color as the seas.

Pluto was considered the smallest planet in the solar system before its status became questioned. Pluto's atmosphere probably contains methane, ammonia, and frozen water. Pluto has 1 satellite. Pluto revolves around the sun every 250 years. Pluto was named after the Roman god of the underworld.

Comets, Asteroids, and Meteors

Astronomers believe that rocky fragments may have been the remains of the birth of the solar system that never formed into a planet. **Asteroids** are found in the region between Mars and Jupiter.

Comets are masses of frozen gases, cosmic dust, and small rocky particles. Astronomers think that most comets originate in a dense comet cloud beyond Pluto. A comet consists of a nucleus, a coma, and a tail. A comet's tail always points away from the sun. The most famous comet, **Halley's Comet,** is named after the person whom first discovered it in 240 B.C. It returns to the skies near earth every 75 to 76 years.

Meteoroids are composed of particles of rock and metal of various sizes. When a meteoroid travels through the earth's atmosphere, friction causes its surface to heat up and it begins to burn. The burning meteoroid falling through the earth's atmosphere is called a **meteor** (also known as a "shooting star").

Meteorites are meteors that strike the earth's surface. A physical example of a meteorite's impact on the earth's surface can be seen in Arizona. The Barringer Crater is a huge meteor crater. There are many other meteor craters throughout the world.

Stars and Other Entities

Astronomers use groups or patterns of stars called **constellations** as reference points to locate other stars in the sky. Familiar constellations include: Ursa Major (also known as the big bear) and Ursa Minor (known as the little bear). Within the Ursa Major, a smaller constellation, the Big Dipper is found. Within the Ursa Minor, a smaller constellation, the Little Dipper is found. Different constellations appear as the earth continues its revolution around the sun with the seasonal changes.

Magnitude stars are 21 of the brightest stars that can be seen from earth. These are the first stars noticed at night. In the Northern Hemisphere there are 15 commonly observed first magnitude stars.

A vast collection of stars are defined as **galaxies**. Galaxies are classified as irregular, elliptical, and spiral. An irregular galaxy has no real structured appearance; most are in their early stages of life. An elliptical galaxy consists of smooth ellipses, containing little dust and gas, but composed of millions or trillion stars. Spiral galaxies are disk-shaped and have extending arms that rotate around its dense center. The solar system is found in the Milky Way galaxy and it is a spiral galaxy.

A **pulsar** is defined as a variable radio source that emits signals in very short, regular bursts; it is believed to be a rotating neutron star.

A **quasar** is defined as an object that photographs like a star but has an extremely large redshift and a variable energy output. It is believed to be the active core of a very distant galaxy.

Black holes are defined as an object that has collapsed to such a degree that light cannot escape from its surface. Light is trapped by the intense gravitational field.

SUBAREA VI.   **INTEGRATION OF KNOWLEDGE
AND UNDERSTANDING OF SCIENCE**

Competency 21.0   **Prepare an organized, developed analysis on a topic
related to one or more of the following: history,
philosophy, and methodology of science; chemistry;
physics; biology; and earth and space science**

The following are three samples of essays on the topic of the scientific process.

A basic sample:

Science is composed of theories, laws, and hypotheses. The first step in
scientific inquiry is posing a question to be answered. Next, a hypothesis is
formed to provide a plausible explanation. An experiment is then proposed and
performed to test this hypothesis. A comparison between the predicted and
observed results is the next step. Conclusions are then formed and it is
determined whether the hypothesis is correct or incorrect. If incorrect, the next
step is to form a new hypothesis and the process is repeated. Science is always
limited by the available technology.

A better sample:

Science is derived from study, observations, and experimentation. Its goal is to
identify and establish principles and theories that may be applied to solve
problems. Scientific theory and experimentation must be repeatable. It is also
possible to disprove or change a theory. Science depends on communication,
agreement, and disagreement among scientists. It is composed of theories, laws,
and hypotheses.

A theory is a principle or relationship that has been verified and accepted through
experiments. A law is an explanation of events that occur with uniformity under
the same conditions. A hypothesis is an educated guess followed by research. A
theory is a proven hypothesis.

Science is limited by the available technology. An example of this would be the
relationship of the discovery of the cell and the invention of the microscope. The
first step in scientific inquiry is posing a question to be answered. Next, a
hypothesis is formed to provide a plausible explanation. An experiment is then
proposed and performed to test this hypothesis. A comparison between the
predicted and observed results is the next step. Conclusions are then formed and
it is determined whether the hypothesis is correct or incorrect. If incorrect, the
next step is to form a new hypothesis and the process is repeated.

An even better sample:

Science may be defined as a body of knowledge that is systematically derived from study, observations, and experimentation. Its goal is to identify and establish principles and theories that may be applied to solve problems. Pseudoscience, on the other hand, is a belief that is not warranted. There is no scientific methodology or application. Some of the more classic examples of pseudoscience include witchcraft, alien encounters or any topic that is explained by hearsay.

Scientific theory and experimentation must be repeatable. It is also possible to be disproved and is capable of change. Science depends on communication, agreement, and disagreement among scientists. It is composed of theories, laws, and hypotheses.

- A theory is the formation of principles or relationships which have been verified and accepted.

- A law is an explanation of events that occur with uniformity under the same conditions (laws of nature, law of gravitation).

- An hypothesis is an unproved theory or educated guess followed by research to best explain a phenomena. A theory is a proven hypothesis.

Science is limited by the available technology. An example of this would be the relationship of the discovery of the cell and the invention of the microscope. As our technology improves, more hypotheses will become theories and possibly laws. Science is also limited by the data that is able to be collected. Data may be interpreted differently on different occasions. Science limitations cause explanations to be changeable as new technologies emerge.

The first step in scientific inquiry is posing a question to be answered. Next, a hypothesis is formed to provide a plausible explanation. An experiment is then proposed and performed to test this hypothesis. A comparison between the predicted and observed results is the next step. Conclusions are then formed and it is determined whether the hypothesis is correct or incorrect. If incorrect, the next step is to form a new hypothesis and the process is repeated.

Sample Test

DIRECTIONS: Read each item and select the best response.

1. **Which of the following is/are true about scientists?**

   I. **Scientists usually work alone.**
   II. **Scientists usually work with other scientists.**
   III. **Scientists achieve more prestige from new discoveries than from replicating established results.**
   IV. **Scientists keep records of their own work, but do not publish it for outside review.**

   (Easy) (Competency 1.0)

   A. I and IV only
   B. II only
   C. II and III only
   D. I and IV only

2. **He was a Monk that through careful control of the pollination of pea plants was able to derive the basic framework of dominant and recessive traits.**
   (Average Rigor)
   (Competency 1.0)

   A. Pasteur
   B. Watson and Crick
   C. Mendel
   D. Mendeleev

3. **A scientific law_____.**
   (Average Rigor)
   (Competency 1.0)

   A. proves scientific accuracy
   B. may never be broken
   C. is the current most accurate explanation for natural phemonon, and experimental data
   D. is the result of one excellent experiment

4. **The most outstanding scientist of the Twentieth Century was known for his theory of:**
   (Average Rigor)
   (Competency 1.0)

   A. laws of motion.
   B. relativity.
   C. continental drift.
   D. evolution.

5. **By discovering the structure of DNA, Watson and Crick made it possible to:**
   (Rigorous)
   (Competency 1.0)

   A. clone DNA.
   B. explain DNA's ability to replicate and control the synthesis of proteins.
   C. sequence human DNA.
   D. predict genetic mutations.

6. Koch's postulates on microbiology include all of the following except:
(Rigorous)
(Competency 1.0)

   A. The same pathogen must be found in every diseased person.
   B. The pathogen must be isolated and grown in culture.
   C. The same pathogen must be isolated from the experimental animal.
   D. Antibodies that react to the pathogen must be found in every diseased person.

7. Which is the correct order of methodology?

   1. collecting data
   2. planning a controlled experiment
   3. drawing a conclusion
   4. hypothesizing a result
   5. re-visiting a hypothesis to answer a question

   (Easy) (Competency 2.0)

   A. 1,2,3,4,5
   B. 4,2,1,3,5
   C. 4,5,1,3,2
   D. 1,3,4,5,2

8. The control group of an experiment is:
(Average Rigor)
(Competency 2.0)

   A. An extra group in which all experimental conditions are the same and the variable being tested is unchanged.
   B. A group of authorities in charge of an experiment.
   C. The group of experimental participants who are given experimental drugs.
   D. A group of subjects that is isolated from all aspects of the experiment.

9. What is the scientific method?
(Average Rigor)
(Competency 2.0)

   A. It is the process of doing an experiment and writing a laboratory report.
   B. It is the process of using open inquiry and repeatable results to establish theories.
   C. It is the process of reinforcing scientific principles by confirming results.
   D. It is the process of recording data and observations.

10. In an experiment measuring the inhibition effect of different antibiotic discs on bacteria grown in Petri dishes, what are the independent and dependent variables respectively? *(Rigorous)* *(Competency 2.0)*

A. number of bacterial colonies and the antibiotic type

B. antibiotic type and the distance between antibiotic and the closest colony

C. antibiotic type and the number of bacterial colonies

D. presence of bacterial colonies and the antibiotic type

11. A scientist exposes mice to cigarette smoke, and notes that their lungs develop tumors. Mice that were not exposed to the smoke do not develop as many tumors. Which of the following conclusions may be drawn from these results?:

I. Cigarette smoke causes lung tumors.

II. Cigarette smoke exposure has a positive correlation with lung tumors in mice.

III. Some mice are predisposed to develop lung tumors.

IV. Cigarette smoke exposure has a positive correlation with lung tumors in humans.

(Rigorous) *(Competency 2.0)*

A. I and II only
B. II only
C. I , II, III and IV
D. II and IV only

12. For her first project of the year, a student is designing a science experiment to test the effects of light and water on plant growth. You should recommend that she _____.

(Rigorous)
(Competency 2.0)

A. manipulate the temperature also
B. manipulate the water pH also
C. determine the relationship between light and water unrelated to plant growth
D. omit either water or light as a variable

13. After an experiment, the scientist states that s/he believes a change in the color of a liquid is due to a change of pH. This is an example of _____ .
(Easy) (Competency 3.0)

A. observing
B. inferring
C. measuring
D. classifying

14. When measuring the volume of water in a graduated cylinder, where does one read the measurement?
(Easy) (Competency 3.0)

A. at the highest point of the liquid
B. at the bottom of the meniscus curve
C. at the closest mark to the top of the liquid
D. at the top of the plastic safety ring

15. In an experiment measuring the growth of bacteria at different temperatures, what is the independent variable?
(Average Rigor)
(Competency 3.0)

A. number of bacteria
B. growth rate of bacteria
C. temperature
D. light intensity

16. When designing a scientific experiment, a student considers all the factors that may influence the results. The process goal is to _____.
(Average Rigor)
(Competency 3.0)

A. recognize and manipulate independent variables
B. recognize and record independent variables
C. recognize and manipulate dependent variables
D. recognize and record dependent variables

17. Which of the following is not an acceptable way for a student to acknowledge sources in a laboratory report?
(Rigorous)
(Competency 3.0)

A. The student tells his/her teacher what sources s/he used to write the report.
B. The student uses footnotes in the text, with sources cited, but not in correct MLA format.
C. The student uses endnotes in the text, with sources cited, in correct MLA format.
D. The student attaches a separate bibliography, noting each use of sources.

18. Which of the following data sets is properly represented by a bar graph?
(Rigorous)
(Competency 3.0)

A. Number of people choosing to buy cars, vs. Color of car bought.
B. Number of people choosing to buy cars, vs. Age of car customer.
C. Number of people choosing to buy cars, vs. Distance from car lot to customer home.
D. Number of people choosing to buy cars, vs. Time since last car purchase.

19. Chemicals should be stored _____.
(Easy) (Competency 4.0)

A. in the principal's office
B. in a dark room
C. in an off-site research facility
D. according to their reactivity with other substances

20. In which situation would a science teacher be legally liable?
     (Average Rigor)
     (Competency 4.0)

     A. The teacher leaves the classroom for a telephone call and a student slips and injures him/herself.
     B. A student removes his/her goggles and gets acid in his/her eye.
     C. A faulty gas line in the classroom causes a fire.
     D. A student cuts him/herself with a dissection scalpel.

21. Accepted procedures for preparing solutions should be made with _____ .
     (Average Rigor)
     (Competency 4.0)

     A. alcohol
     B. hydrochloric acid
     C. distilled water
     D. tap water

22. Which is the most desirable tool to use to heat substances in a middle school laboratory?
     (Average Rigor)
     (Competency 4.0)

     A. alcohol burner
     B. freestanding gas burner
     C. bunsen burner
     D. hot plate

23. Electrophoresis uses electrical charges of molecules to separate them according to their:
     (Rigorous)
     (Competency 4.0)

     A. polarization.
     B. size.
     C. type.
     D. shape.

24. Experiments may be done with any of the following animals except _____ .
     (Rigorous)
     (Competency 4.0)

     A. birds
     B. invertebrates
     C. lower order life
     D. frogs

25. In a science experiment, a student needs to repeatly dispense very small measured amounts of liquid into a well mixed solution. Which of the following is the best choice for his/her equipment to use?
     (Rigorous)
     (Competency 4.0)

     A. pipette, stirring rod, beaker
     B. burette with burette stand, stir plate, beaker
     C. volumetric flask, dropper, stirring rod
     D. beaker, graduated cylinder, stir plate

26. **The elements in the modern periodic table are arranged** _____ .
(Easy) (Competency 5.0)

 A. in numerical order by atomic number
 B. randomly
 C. in alphabetical order by chemical symbol
 D. in numerical order by atomic mass

27. **Which parts of an atom are located inside the nucleus?**
(Easy) (Competency 5.0)

 A. electrons and neutrons
 B. protons and neutrons
 C. protons only
 D. neutrons only

28. **What is specific gravity?**
(Average Rigor)
(Competency 5.0)

 A. the mass of an object
 B. the ratio of the density of a substance to the density of water
 C. density
 D. the pull of the earth's gravity on an object

29. **The electrons in an atom that are used to form a chemical bond are called** _____ .
(Average Rigor)
(Competency 5.0)

 A. outer shell electrons
 B. excited electrons
 C. valence electrons
 D. reactive electrons

30. **The chemical equation for water formation is: $2H_2 + O_2 \rightarrow 2H_2O$. Which of the following is an incorrect interpretation of this equation?**
(Rigorous)
(Competency 5.0)

 A. Two moles of hydrogen gas and one mole of oxygen gas combine to make two moles of water.
 B. Two grams of hydrogen gas and one gram of oxygen gas combine to make two grams of water.
 C. Two molecules of hydrogen gas and one molecule of oxygen gas combine to make two molecules of water.
 D. Four atoms of hydrogen (combined as a diatomic gas) and two atoms of oxygen (combined as a diatomic gas) combine to make two molecules of water.

31. Which of the following is not a property of metalloids?
(Rigorous) (Competency 5.0)

A. Metalloids are solids at standard temperature and pressure.
B. Metalloids can conduct electricity to a limited extent.
C. Metalloids are found in groups 13 through 17.
D. Metalloids all favor ionic bonding.

32. Matter's phase (solid, liquid, or gas) is identified by its:
(Easy) (Competency 6.0)

A. color and size.
B. shape and volume.
C. size and volume.
D. color and volume.

33. When heat is added to most solids, they expand. Why is this the case?
(Average Rigor) (Competency 6.0)

A. The molecules get bigger.
B. The faster molecular motion leads to greater distance between the molecules.
C. The molecules develop greater repelling electric forces.
D. The molecules form a more rigid structure.

34. Which of the following is most accurate?
(Average Rigor) (Competency 6.0)

A. Mass is always constant; weight may vary by location.
B. Mass and weight are both always constant.
C. Weight is always constant; mass may vary by location.
D. Mass and weight may both vary by location.

35. The change in phase from liquid to gas is called:
(Rigorous) (Competency 6.0)

A. evaporation.
B. condensation.
C. vaporization.
D. boiling.

36. A seltzer tablet changing into bubbles is an example of:
(Rigorous) (Competency 6.0)

A. a physical change
B. a chemical change
C. conversion
D. diffusion

37. **Energy is:**
    (Rigorous)
    (Competency 6.0)

    A. The combination of power and work.
    B. The ability to cause change in matter.
    C. The transfer of power when force is applied to a body.
    D. Physical force.

38. **Catalysts assist reactions by _____ .**
    (Easy) (Competency 7.0)

    A. lowering required activation energy
    B. maintaining precise pH levels
    C. keeping systems at equilibrium
    D. changing the starting amounts of reactants

39. **What is the best explanation for isomerization?**
    (Average Rigor)
    (Competency 7.0)

    A. A chemical reaction in which a molecule changes shape, but no atoms are lost or gained.
    B. A chemical reaction in which a molecule changes shape, but one or more atoms are lost.
    C. A chemical reaction in which a molecule changes shape, but one or more atoms are gained.
    D. A chemical reaction in which a compound is broke down into its constitute parts.

40. **Which of the following is a correct definition for 'chemical equilibrium'?**
(Average Rigor)
(Competency 7.0)

    A. Chemical equilibrium is when the forward and backward reaction rates are equal. The reaction may continue to proceed forward and backward.

    B. Chemical equilibrium is when the forward and backward reaction rates are equal, and equal to zero. The reaction does not continue.

    C. Chemical equilibrium is when there are equal quantities of reactants and products.

    D. Chemical equilibrium is when acids and bases neutralize each other fully.

41. **Which of the following will not change in a chemical reaction?**
(Average Rigor)
(Competency 7.0)

    A. number of moles of products

    B. atomic number of one of the reactants

    C. mass (in grams) of one of the reactants

    D. rate of reaction

42. **Carbon bonds with hydrogen by _____ .**
(Rigorous)
(Competency 7.0)

    A. ionic bonding

    B. non-polar covalent bonding

    C. polar covalent bonding

    D. strong nuclear force

43. **Which reaction below is a decomposition reaction?**
(Rigorous)
(Competency 7.0)

A. $HCl + NaOH \rightarrow NaCl + H_2O$
B. $C + O_2 \rightarrow CO_2$
C. $2H_2O \rightarrow 2H_2 + O2$
D. $CuSO_4 + Fe \rightarrow FeSO_4 + Cu$

44. **A cup of hot liquid and a cup of cold liquid are both sitting in a room at a temperature of 72 degrees Fahrenheit and 25% humidity. Both cups are thin plastic. Which of the following is a true statement?**
(Easy) (Competency 8.0)

A. There will be condensation on the outside of both cups.
B. There will be condensation on the outside of the hot liquid cup, but not on the cold liquid cup.
C. There will be condensation on the outside of the cold liquid cup, but not on the hot liquid cup.
D. There will not be condensation on the outside of either cup.

45. **If the volume of a confined gas is increased, what happens to the pressure of the gas? You may assume that the gas behaves ideally, and that temperature and number of gas molecules remain constant.**
(Average Rigor)
(Competency 8.0)

A. The pressure increases.
B. The pressure decreases.
C. The pressure stays the same.
D. There is not enough information given to answer this question.

46. **What is necessary for ion diffusion to occur spontaneously? (Diffusion relates to the movement of particles.)**
(Average Rigor)
(Competency 8.0)

A. carrier proteins
B. energy from an outside source
C. a concentration gradient
D. activation energy

47. The relationships between pressure and temperature and between temperature and volume in a gas are examples of:
(Average Rigor)
(Competency 8.0)

    A.   indirect variations.
    B.   direct variations.
    C.   Boyle's law.
    D.   Charles' law.

48. Which of the following is not true about phase change in matter?
(Rigorous)
(Competency 8.0)

    A.   Solid water and liquid ice can coexist at water's freezing point.
    B.   At 7 degrees Celsius, water is always in liquid phase.
    C.   Matter changes phase when enough energy is gained or lost.
    D.   Different phases of matter are characterized by differences in molecular motion.

49. Pressure is measured in a unit called the pascal. One pascal is equal to one _____ of force pushing on one square meter of area.
(Rigorous)
(Competency 8.0)

    A.   joule
    B.   calorie
    C.   newton
    D.   erg

50. All of the following are considered Newton's laws except for:
(Easy) (Competency 9.0)

    A.   An object in motion will continue in motion unless acted upon by an outside force.
    B.   For every action force, there is an equal and opposite reaction force.
    C.   Nature abhors a vacuum.
    D.   Mass can be considered the ratio of force to acceleration.

51. A ball rolls down a smooth hill. You may ignore air resistance. Which of the following is a true statement?
(Average Rigor)
(Competency 9.0)

   A. The ball has more energy at the start of its descent than just before it hits the bottom of the hill, because it is higher up at the beginning.
   B. The ball has less energy at the start of its descent than just before it hits the bottom of the hill, because it is moving more quickly at the end.
   C. The ball has the same energy throughout its descent, because positional energy is converted to energy of motion.
   D. The ball has the same energy throughout its descent, because a single object (such as a ball) cannot gain or lose energy.

52. The force of gravity on earth causes all bodies in free fall to _____ .
(Average Rigor)
(Competency 9.0)

   A. fall at the same speed
   B. accelerate at the same rate
   C. reach the same terminal velocity
   D. move in the same direction

53. Dynamics is the study of the relationship between:
(Rigorous)
(Competency 9.0)

   A. heat and energy
   B. heat and motion
   C. motion and work
   D. motion and force

54. Since ancient times, people have been entranced with bird flight. What is the key to bird flight?
    (Rigorous)
    (Competency 9.0)

    A. Bird wings are a particular shape and composition, that causes the air flow over the wing to travel faster than the air flow under the wing.
    B. Birds flap their wings quickly enough to propel themselves.
    C. Bird wings are a particular shape and composition, that causes the air flow under the wing to travel faster than the air flow over the wing.
    D. Birds flapping of their wings creates a downward force that opposes gravity.

55. Newton's laws are taught in science classes because _____.
    (Rigorous)
    (Competency 9.0)

    A. they are the correct analysis of inertia, gravity, and forces
    B. they are a close approximation to correct physics, for usual conditions on earth
    C. they accurately incorporate relativity into studies of forces
    D. Newton was a well-respected scientist in his time

56. The transfer of heat by electromagnetic waves is called _____ .
    (Easy) (Competency 10.0)

    A. conduction
    B. convection
    C. phase change
    D. radiation

57. A long silver bar has a temperature of 50 degrees Celsius at one end and 0 degrees Celsius at the other end. The bar will reach thermal equilibrium (barring outside influence) by the process of heat

_____.

(Average Rigor)
(Competency 10.0)

  A. conduction
  B. convection
  C. radiation
  D. phase change

58. The law of conservation of energy states that:
(Average Rigor)
(Competency 10.0)

  A. There must be the same number of products and reactants in any chemical equation.
  B. Mass and energy can be interchanged.
  C. Energy is neither created nor destroyed, but may change form.
  D. One form energy must remain intact (or conserved) in all reactions.

59. When you step out of the shower, the floor feels colder on your feet than the bathmat. Which of the following is the correct explanation for this phenomenon?
(Rigorous)
(Competency 10.0)

  A. The floor is colder than the bathmat.
  B. The bathmat being smaller that the floor quickly reaches equilibrium with your body temperature.
  C. Heat is conducted more easily into the floor.
  D. Water is absorbed from your feet into the bathmat so it doesn't evaporate as quickly as it does off the floor thus not cooling the bathmat as quickly.

60. **What is the best explanation of the term "latent heat"?**
(Rigorous)
(Competency 10.0)

A. The amount of heat it takes to change a solid to a liquid.
B. The amount of heat being radiated by an object.
C. The amount heat needed to change a substance to undergo a phase change.
D. The amount of heat it takes to change a liquid to a gas.

61. **Sound waves are produced by _____ .**
(Easy) (Competency 11.0)

A. pitch
B. noise
C. vibrations
D. sonar

62. **Sound can be transmitted in all of the following except _____ .**
(Easy) (Competency 11.0)

A. air
B. water
C. A diamond
D. a vacuum

63. **The Doppler effect is associated most closely with which property of waves?**
(Average Rigor)
(Competency 11.0)

A. amplitude
B. wavelength
C. frequency
D. intensity

64. **The speed of light is different in different materials. This is responsible for _____ .**
(Average Rigor)
(Competency 11.0)

A. interference
B. refraction
C. reflection
D. relativity

65. **A converging lens produces a real image _____.**
(Rigorous)
(Competency 11.0)

 A.   always
 B.   never
 C.   when the object is within one focal length of the lens
 D.   when the object is further than one focal length from the lens

66. **As a train approaches, the whistle sounds _____.**
(Rigorous)
(Competency 11.0)

 A.   higher, because it has a higher apparent frequency
 B.   lower, because it has a lower apparent frequency
 C.   higher, because it has a lower apparent frequency
 D.   lower, because it has a higher apparent frequency

67. **The electromagnetic radiation with the longest wavelength is/are _____.**
(Easy) (Competency 12.0)

 A.   radio waves
 B.   red light
 C.   X-rays
 D.   ultraviolet light

68. **In Ohm's Law (I = V/R), the V represents:**
(Average Rigor)
(Competency 12.0)

 A.   current.
 B.   amperes.
 C.   potential difference.
 D.   resistance.

69. **Which of the following is not a characteristic of all electrically charged objects?**
(Average Rigor)
(Competency 12.0)

 A.   Opposites attract.
 B.   Like repels like.
 C.   Charge is conserved.
 D.   A magnetic charge develops.

70. **Resistance is measured in units called _____.**
(Average Rigor)
(Competency 12.0)

 A.   watts
 B.   volts
 C.   ohms
 D.   current

71. A light bulb is connected in series with a rotating coil within a magnetic field. The brightness of the light may be increased by any of the following except:
(Rigorous)
(Competency 12.0)

    A. Rotating the coil more rapidly.
    B. Using more loops in the coil.
    C. Using a different color wire for the coil.
    D. Using a stronger magnetic field.

72. A 10 ohm resistor and a 50 ohm resistor are connected in parallel. If the current in the 10 ohm resistor is 5 amperes, the current (in amperes) running through the 50 ohm resistor is
(Rigorous)
(Competency 12.0)

    A. 1.
    B. 50.
    C. 25.
    D. 60.

73. Which is the correct sequence of insect development?
(Easy) (Competency 13.0)

    A. Egg, pupa, larva, adult.
    B. Egg, larva, pupa, adult.
    C. Egg, adult, larva, pupa.
    D. Pupa, egg, larva, adult.

74. Which part of a plant is responsible for transporting water?
(Easy) (Competency 13.0)

    A. phloem
    B. xylem
    C. stomata
    D. cortex

75. Animals with a notochord or backbone are in the phylum _____.
(Average Rigor)
(Competency 13.0)

    A. arthropoda
    B. chordata
    C. mollusca
    D. ammalia.

76. **What cell organelle contains the cell's stored food?**
(Rigorous)
(Competency 13.0)

A. vacuoles
B. Golgi apparatus
C. ribosomes
D. lysosomes

77. **Laboratory researchers have classified fungi as distinct from plants because the cell walls of fungi _____ .**
(Rigorous)
(Competency 13.0)

A. contain chitin
B. contain yeast
C. are more solid
D. are less solid

78. **A product of anaerobic respiration in animals is**

_____.
(Rigorous)
(Competency 13.0)

A. carbon dioxide
B. lactic acid
C. oxygen
D. sodium chloride

79. **In the law of dominance:**
(Easy) (Competency 14.0)

A. Only one of the two possible alleles from each parent is passed on to the offspring.
B. Alleles sort independently of each other.
C. One trait may cover up the allele of the other trait.
D. Flowers have white alleles and purple alleles.

80. **Which process(es) result(s) in a haploid chromosome number?**
(Average Rigor)
(Competency 14.0)

A. mitosis
B. meiosis
C. both mitosis and meiosis
D. neither mitosis nor meiosis

81. **A white flower is crossed with a red flower. Which of the following is a sign of incomplete dominance?**
(Average Rigor)
(Competency 14.0)

A. pink flowers
B. red flowers
C. white flowers
D. no flowers

82. A child has type O blood. Her father has type A blood, and her mother has type B blood. What are the genotypes of the father and mother, respectively?
(Average Rigor)
(Competency 14.0)

A. AO and BO
B. AA and AB
C. OO and BO
D. AO and BB

83. A monohybrid cross has four possible gene combinations. How many gene combinations are possible in a dihybrid cross?
(Rigorous)
(Competency 14.0)

A. eight
B. sixteen
C. thirty-two
D. sixty-four

84. An Arabic horse's purebred bloodline makes it a good example of:
(Rigorous)
(Competency 14.0)

A. a homozygous animal.
B. a heterozygous animal.
C. codominance.
D. A polygenic haracter.

85. A duck's webbed feet are examples of _____ .
(Easy) (Competency 15.0)

A. mimicry
B. structural adaptation
C. protective resemblance
D. protective coloration

86. An animal choosing its mate because of attractive plumage or a strong mating call is an example of:
(Average Rigor)
(Competency 15.0)

A. sexual selection.
B. natural selection.
C. peer selection.
D. linkage.

87. Which of the following is not considered to be a cause of evolution?
(Average Rigor)
(Competency 15.0)

A. sexual reproduction
B. immigration
C. large populations
D. random mating

88. **Which of the following is not one of the principles of Darwin's theory of natural selection?**
(Average Rigor)
(Competency 15.0)

A. More individuals are produced than will survive.

B. The individuals in a certain species vary from generation to generation.

C. Only the fittest members of a species survive.

D. Some genes allow for better survival of an animal.

89. **Which of the following is the best example of an explanation of the theory of evolution?**
(Rigorous)
(Competency 15.0)

A. Giraffes need to reach higher for leaves to eat, so their necks stretch. The giraffe babies are then born with longer necks. Eventually, there are more long-necked giraffes in the population.

B. Giraffes with longer necks are able to reach more leaves, so they eat more and have more babies than other giraffes. Eventually, there are more long-necked giraffes in the population.

C. Giraffes want to reach higher for leaves to eat, so they release enzymes into their bloodstream, which in turn causes fetal development of longer-necked giraffes. Eventually, there are more long-necked giraffes in the population.

D. Giraffes with long necks are more attractive to other giraffes, so they get the best mating partners and have more babies. Eventually, there are more long-necked giraffes in the population.

90. What is the principle driving force for evolution of antibiotic resistant bacteria?
(Rigorous)
(Competency 15.0)

   A. mutation
   B. reproduction method
   C. population size
   D. emigration

91. What are the most significant and prevalent elements in the biosphere? (The biosphere contains all biomes.)
(Easy) (Competency 16.0)

   A. carbon, hydrogen, oxygen, nitrogen, phosphorus
   B. carbon, hydrogen, sodium, iron, calcium
   C. carbon, oxygen, sulfur, manganese, iron
   D. carbon, hydrogen, oxygen, nickel, sodium, nitrogen

92. Which of the following is the most accurate definition of a nonrenewable resource?
(Average Rigor)
(Competency 16.0)

   A. A nonrenewable resource is never replaced once used.
   B. A nonrenewable resource is replaced on a timescale that is very long relative to human life spans.
   C. A nonrenewable resource is a resource that can only be manufactured by humans.
   D. A nonrenewable resource is a species that has already become extinct.

93. A wrasse (fish) cleans the teeth of other fish by eating away plaque. This is an example of _____ between the fish.
(Average Rigor)
(Competency 16.0)

   A. parasitism
   B. symbiosis (mutualism)
   C. competition
   D. predation

94. Which of the following terms does not describe a way that the human race has had a negative impact on the biosphere?
(Rigorous)
(Competency 16.0)

    A.   biological magnification
    B.   pollution
    C.   carrying capacity
    D.   simplifcation of the food web

95. Which one of the following biomes makes up the greatest percentage of the biosphere?
(Rigorous)
(Competency 16.0)

    A.   desert
    B.   tropical rain forest
    C.   marine
    D.   temperate deciduous forest

96. In commensalism:
(Rigorous)
(Competency 16.0)

    A.   Two species occupy a simlar place; one species benefits from the other and one species is harmed by the other.
    B.   Two species occupy a similar place and neither is harmed or benefits.
    C.   Two species occupy the similar place and both species benefit.
    D.   Two species occupy the same habitat and one preys upon the other.

97. When water falls to a cave floor and evaporates, it may deposit calcium carbonate. This process leads to the formation of which of the following?
(Easy) (Competency 17.0)

    A.   stalactites
    B.   stalagmites
    C.   fault lines
    D.   sedimentary rocks

98. Fossils are usually found in _____ rock.
(Easy) (Competency 17.0)

    A.   igneous
    B.   sedimentary
    C.   metamorphic
    D.   cumulus

99. Which of the following is the longest (largest) unit of geological time?
(Average Rigor)
(Competency 17.0)

A. solar year
B. epoch
C. period
D. era

100. A contour line that has tiny comb-like lines along the inner edge indicates a ___?
(Average Rigor)
(Competency 17.0)

A. depression
B. mountain
C. valley
D. river

101. Which of the following is a type of igneous rocks?
(Rigorous)
(Competency 17.0)

A. quartz
B. granite
C. obsidian
D. all of the above

102. Lithification refers to one process to create

_____.

(Rigorous)
(Competency 17.0)

A. metamorphic rocks
B. sedimentary rocks
C. igneous rocks
D. lithium oxide

103. Which of the following is the best explanation of the fundamental concept of uniformitarianism?
(Rigorous)
(Competency 17.0)

A. The types and varieties of life reveal a uniform progression over time.
B. The physical, chemical and biological laws that operate in the geologic past operate in the same way today.
C. Debris from catastrophic events (i.e., volcanoes and meteorites) will be evenly distributed over the effected area.
D. The frequency and intensity of major geologic events will remain consistent over long periods of time.

104. The salinity of ocean water is closest to _____.
(Easy) (Competency 18.0)

A. 0.035 %
B. 0.5 %
C. 3.5 %
D. 15 %

105. The theory of 'sea floor spreading' explains _____.
(*Average Rigor*)
(*Competency 18.0*)

   A. the shapes of the continents
   B. how continents collide
   C. how continents move apart
   D. how continents sink to become part of the ocean floor

106. What is the most accurate description of the water cycle?
(*Average Rigor*)
(*Competency 18.0*)

   A. Rain comes from clouds, filling the ocean. The water then evaporates and becomes clouds again.
   B. Water circulates from rivers into groundwater and back, while water vapor circulates in the atmosphere.
   C. Water is conserved except for chemical or nuclear reactions, and any drop of water could circulate through clouds, rain, ground-water, and surface-water.
   D. Weather systems cause chemical reactions to break water into its atoms.

107. What is the source for most of the United States' drinking water?
(*Rigorous*)
(*Competency 18.0*)

   A. desalinated ocean water
   B. surface water (lakes, streams, mountain runoff)
   C. rainfall into municipal reservoirs
   D. groundwater

108. Surface ocean currents are caused by which of the following?
(*Rigorous*)
(*Competency 18.0*)

   A. temperature
   B. density changes in water
   C. wind
   D. tidal forces

109. Mount Kilauea on the island of Hawaii is a very active volcano that has continuous lava flow into the ocean near it. What is the name of the type of shoreline created at the point where the lava flows meet the water?
(*Rigorous*)
(*Competency 18.0*)

   A. stacking
   B. submerged
   C. developing
   D. emergent

110. Which of the following instruments measures wind speed?
(Easy) (Competency 19.0)

   A. barometer
   B. anemometer
   C. thermometer
   D. weather vane

111. The calm point at the center of a storm such as a hurricane is often called the "eye" of the storm. This "eye" is caused by:
(Average Rigor) (Competency 19.0)

   A. centrepidal force.
   B. a high-pressure air mass.
   C. a low-pressure air mass.
   D. heavier precipitation in the area.

112. Air masses moving toward or away form the Earth's surface is called ____.
(Average Rigor) (Competency 19.0)

   A. wind
   B. breeze
   C. air currents
   D. doldrums

113. Which type of cloud is most likely to produce precipitation?
(Average Rigor) (Competency 19.0)

   A. cirrocumulus
   B. stratocumulus
   C. cumulonimbus
   D. cirrostratus

114. Air moving northward from the horse latitudes produces a belt of winds called the ____.
(Rigorous) (Competency 19.0)

   A. prevailing westerlies
   B. north westerlies
   C. trade winds
   D. prevailing easterlies

115. Which is a form of precipitation?
(Rigorous) (Competency 19.0)

   A. snow
   B. frost
   C. fog
   D. all of the above

116. Which of the following units is not a measure of distance?
(Easy) (Competency 20.0)

   A. AU (astronomical unit)
   B. light year
   C. parsec
   D. lunar year

117. A star's brightness is referred to as _____.
(Average Rigor)
(Competency 20.0)

    A. magnitude
    B. mass
    C. apparent magnitude
    D. Intensity

118. Which of the following is the best definition for "meteroid"?
(Average Rigor)
(Competency 20.0)

    A. A meteroid is material from outer space, that is composed of particles of rock and metal.
    B. A meteroid is material from outer space, that has struck the earth's surface.
    C. A meteroid is an element that has properties of both metals and nonmetals.
    D. A meteroid is a very small unit of length measurement.

119. A telescope that collects light by using a concave mirror and can produce small images is called a

_____.
(Average Rigor)
(Competency 20.0)

    A. radioactive telescope
    B. reflecting telescope
    C. refracting telescope
    D. optical telescope

120. **What is the main difference between the "condensation hypothesis" and the "tidal hypothesis" for the origin of the solar system?**
(Rigorous)
(Competency 20.0)

A. The tidal hypothesis can be tested, but the condensation hypothesis cannot.

B. The tidal hypothesis proposes a near collision of two stars pulling on each other, but the condensation hypothesis proposes condensation of rotating clouds of dust and gas.

C. The tidal hypothesis explains how tides began on planets such as earth, but the condensation hypothesis explains how water vapor became liquid on earth.

D. The tidal hypothesis is based on Aristotelian physics, but the condensation hypothesis is based on Newtonian mechanics.

121. **The planet with true retrograde rotation is**
_____.
(Rigorous)
(Competency 20.0)

A. Pluto
B. Uranus
C. Venus
D. Saturn

Answer Key

1.	C	26.	A	51.	C	76.	A	101.	D
2.	C	27.	B	52.	B	77.	A	102.	B
3.	C	28.	B	53.	D	78.	B	103.	B
4.	B	29.	C	54.	A	79.	C	104.	C
5.	B	30.	B	55.	B	80.	B	105.	C
6.	D	31.	D	56.	D	81.	A	106.	C
7.	B	32.	B	57.	A	82.	A	107.	D
8.	A	33.	B	58.	C	83.	B	108.	C
9.	B	34.	A	59.	C	84.	A	109.	D
10.	B	35.	A	60.	C	85.	B	110.	B
11.	B	36.	B	61.	C	86.	A	111.	C
12.	D	37.	B	62.	D	87.	D	112.	C
13.	B	38.	A	63.	C	88.	C	113.	C
14.	B	39.	A	64.	C	89.	B	114.	A
15.	C	40.	A	65.	D	90.	C	115.	A
16.	A	41.	B	66.	A	91.	A	116.	D
17.	A	42.	C	67.	A	92.	B	117.	A
18.	A	43.	C	68.	C	93.	B	118.	A
19.	D	44.	C	69.	D	94.	C	119.	B
20.	A	45.	B	70.	C	95.	C	120.	B
21.	C	46.	C	71.	C	96.	B	121.	C
22.	D	47.	B	72.	A	97.	B		
23.	B	48.	B	73.	B	98.	B		
24.	A	49.	C	74.	B	99.	D		
25.	B	50.	C	75.	B	100.	A		

Rigor Table

	Easy %20	Average Rigor %40	Rigorous %40
Question #	1, 7 ,13, 14, 19, 26, 27, 32, 38, 45, 51, 58, 64, 65, 70, 76, 77, 83, 89, 95, 101, 102, 108, 114, 120	2, 3, 4, 8, 9, 15, 16, 20, 21, 22, 28, 29, 33, 34, 39, 40, 41, 46, 47, 48, 52, 54, 59, 60, 66, 67, 71, 72, 73, 78, 84, 85, 86, 90, 91, 92, 96, 97, 103, 104, 109, 110, 115, 116, 117, 121, 122, 123	5, 6, 10, 11, 12, 17, 18, 23, 24, 25 30, 31, 35, 36, 37, 43, 44, 49, 50 ,55, 56, 57, 61, 62, 68, 69, 74, 75, 80, 81, 82, 87, 88, 93, 94, 98, 99, 100, 105, 106, 107, 111, 112, 113, 118, 119, 124, 125

Rationales with Sample Questions

1. **Which of the following is/are true about scientists?**
    I.   **Scientists usually work alone.**
    II. **Scientists usually work with other scientists.**
    III. **Scientists achieve more prestige from new discoveries than from replicating established results.**
    IV. **Scientists keep records of their own work, but do not publish it for outside review.**

    (Easy) (Competency 1.0)

A   I and IV only
B   II only
C   II and III only
D   I and IV only

Answer: C. II and III only

In the scientific community, scientists nearly always work in teams, both within their institutions and across several institutions. This eliminates (I) and requires (II), leaving only answer choices (B) and (C). Scientists do achieve greater prestige from new discoveries, so the answer must be (C). Note that scientists must publish their work in peer-reviewed journals, eliminating (IV) in any case.

2. **He was a Monk that through careful control of the pollination of pea plants was able to derive the basic framework of dominant and recessive traits.**
    (Average Rigor) (Competency 1.0)

A   Pasteur
B   Watson and Crick
C   Mendel
D   Mendeleev

Answer: C. Mendel.

Gregor Mendel was a ninteenth-century Austrian botanist, who derived "laws" governing inherited traits. His work led to the understanding of dominant and recessive traits, carried by biological markers. Mendel crossbred different kinds of pea plants with varying features and observed the resulting new plants. He showed that genetic characteristics are not passed identically from one generation to the next. (Pasteur, Watson, Crick, and Mendeleev were other scientists with different specialties.) This is consistent only with answer (C).

3.   A scientific law_____.
      (Average Rigor) (Competency 1.0)

A   proves scientific accuracy
B   may never be broken
C   is the current most accurate explanation for natural phemonon, and experimental data
D   is the result of one excellent experiment

Answer: C. is the current most accuracte explanation for natural phemonon

A scientific law is a tool that is used for making predictions about natural occurances and experimental data. A scientific law is always the result of many experiments, and is implied or supported by various results. Therefore, such a law may be revised in light of new data, and may be replaced by a new law that is a more accurate. For example Newton's laws have been replaced by Einstein's's relativity, because the realitivity can explain natural occurances that Newton's laws cannot.   Therefore, the answer must be (C).

4.   The most outstanding scientist of the Twentieth Century was known for his theory of:
      (Average Rigor) (Competency 1.0)

A   laws of motion.
B   relativity.
C   continental drift.
D   evolution.

Answer: B. Relativity

Newton's laws of motion were developed during the nineteenth century. While Wegener's theory of continental drift expanded our geologic horizons, and Darwin's evolutionary theories have become the subject of much study and debate, it was Einstein's theory of relativity that was the most uncontestedly outstanding discovery of the twentieth century. Thus, the answer is (B).

5.   By discovering the structure of DNA, Watson and Crick made it
      possible to:
      (Rigorous) (Competency 1.0)

A   clone DNA.
B   explain DNA's ability to replicate and control the synthesis of proteins.
C   sequence human DNA.
D   predict genetic mutations.

Answer: B. explain DNA's ability to replicate and control the synthesis of proteins.

While more recent discoveries have made it possible to sequence the human genome, clone DNA, and predict genetic mutations, it was Watson's and Crick's discovery of DNA's structure that made it possible for scientists to understand and therefore explain DNA's ability to replicate and control the synthesis of proteins. Thus, the correct answer is (B).

6.   Koch's postulates on microbiology include all of the following
      except:
      (Rigorous) (Competency 1.0)

A   The same pathogen must be found in every diseased person.
B   The pathogen must be isolated and grown in culture.
C   The same pathogen must be isolated from the experimental animal.
D   Antibodies that react to the pathogen must be found in every
      diseased person.

Answer: D. Antibodies that react to the pathogen must be found in every diseased person.

Koch postulated that the same pathogen must be found in every diseased person, this pathogen must be isolated and grown in a culture, the disease must be induced in experimental animals from the same pathogen, and that this pathogen must then be isolated from the experimental animal. Koch's postulates do not mention antibodies being found. Thus, the answer is (D).

7.   **Which is the correct order of methodology?**
      1.   **collecting data**
      2.   **planning a controlled experiment**
      3.   **drawing a conclusion**
      4.   **hypothesizing a result**
      5.   **re-visiting a hypothesis to answer a question**
      (Easy) (Competency 2.0)

A   1,2,3,4,5
B   4,2,1,3,5
C   4,5,1,3,2
D   1,3,4,5,2

Answer: B. 4, 2, 1, 3, 5

The correct methodology for the scientific method is first to make a meaningful hypothesis (educated guess), then plan and execute a controlled experiment to test that hypothesis. Using the data collected in that experiment, the scientist then draws conclusions and attempts to answer the original question related to the hypothesis. This is consistent only with answer (B).

8.   **The control group of an experiment is:**
      (Average Rigor) (Competency 2.0)

A   An extra group in which all experimental conditions are the same and the variable being tested is unchanged.
B   A group of authorities in charge of an experiment.
C   The group of experimental participants who are given experimental drugs.
D   A group of subjects that is isolated from all aspects of the experiment.

Answer: A. An extra group in which all experimental conditions are the same.

A group of authorities in charge of an experiment, while they might be in control, they are not a control group. The group of experimental participants given the experimental drugs would be the experimental group, and a group of subjects isolated from all aspects of the experiment would not be part of the experiment at all. Thus, the answer is (A).

9.   **What is the scientific method?**
      (Average Rigor) (Competency 2.0)

A   It is the process of doing an experiment and writing a laboratory report.
B   It is the process of using open inquiry and repeatable results to establish theories.
C   It is the process of reinforcing scientific principles by confirming results.
D   It is the process of recording data and observations.

Answer: B. It is the process of using open inquiry and repeatable results to establish theories.

Scientific research often includes elements from answers (A), (C), and (D), but the basic underlying principle of the scientific method is that people ask questions and do repeatable experiments to answer those questions and develop informed theories of why and how things happen. Therefore, the best answer is (B).

10.   **In an experiment measuring the inhibition effect of different antibiotic discs on bacteria grown in Petri dishes, what are the independent and dependent variables respectively?**
      (Rigorous) (Competency 2.0)

A   number of bacterial colonies and the antibiotic type
B   antibiotic type and the distance between antibiotic and the closest colony
C   antibiotic type and the number of bacterial colonies
D   presence of bacterial colonies and the antibiotic type

Answer: B. antibiotic type and the distance between antibiotic and the closest colony

To answer this question, recall that the independent variable in an experiment is the entity that is changed by the scientist, in order to observe the effects on the dependent variable. In this experiment, antibiotic used is purposely changed so it is the independent variable. Answers A and D list antibiotic type as the dependent variable and thus cannot be the answer, leaving answers B and C as the only two viable choices. The best answer is B, because it measures at what concentration of the antibiotic the bacteria are able to grow at, (as you move from the source of the antibiotic the concentration decreases). Answer C is not as effective because it could be interpreted that that a plate that shows a large number of colonies a greater distance from the antibiotic is a less effective antibiotic than a plate a smaller number of colonies in close proximity to the antibiotic disc, which is reverse of the actually result.

11.   A scientist exposes mice to cigarette smoke, and notes that their
       lungs develop tumors. Mice that were not exposed to the smoke do
       not develop as many tumors. Which of the following conclusions
       may be drawn from these results?:
       I.   Cigarette smoke causes lung tumors.
       II.   Cigarette smoke exposure has a positive correlation with lung
              tumors in mice.
       III.   Some mice are predisposed to develop lung tumors.
       IV.   Cigarette smoke exposure has a positive correlation with lung
              tumors in humans.
       (Rigorous) (Competency 2.0)

A   I and II only
B   II only
C   I , II, III and IV
D   II and IV only

Answer: B. II only

Although cigarette smoke has been found to cause lung tumors (and many other
problems), this particular experiment shows only that there is a positive
correlation between smoke exposure and tumor development in these mice. It
may be true that some mice are more likely to develop tumors than others, which
is why a control group of identical mice should have been used for comparison.
Mice are often used to model human reactions, but this is as much due to their
low financial and emotional cost as it is due to their being a "good model" for
humans, and thus this scientist cannot make the conclusion that cigarette smoke
exposure has a positive correlation with lung tumors in humans based on this
data alone. Therefore, the answer must be (B).

12. **For her first project of the year, a student is designing a science experiment to test the effects of light and water on plant growth. You should recommend that she _____.**
   (Rigorous) (Competency 2.0)

A   manipulate the temperature also
B   manipulate the water pH also
C   determine the relationship between light and water unrelated to plant growth
D   omit either water or light as a variable

Answer: D. omit either water or light as a variable

As a science teacher for middle-school-aged kids, it is important to reinforce the idea of 'constant' vs. 'variable' in science experiments. At this level, it is wisest to have only one variable examined in each science experiment. (Later, students can hold different variables constant while investigating others.) Therefore it is counterproductive to add in other variables (answers (A)or (B)). It is also irrelevant to determine the light-water interactions aside from plant growth (C). So the only possible answer is (D).

13. **After an experiment, the scientist states that s/he believes a change in the color of a liquid is due to a change of pH. This is an example of _____.**
   (Easy) (Competency 3.0)

A   observing
B   inferring
C   measuring
D   classifying

Answer: B. inferring

To answer this question, note that the scientist has observed a change in color, and has then made a guess as to its reason. This is an example of inferring. The scientist has not measured or classified in this case. Although s/he has observed [the color change], the explanation of this observation is inferring (B).

14.	When measuring the volume of water in a graduated cylinder, where does one read the measurement?
	(Easy) (Competency 3.0)

A	at the highest point of the liquid
B	at the bottom of the meniscus curve
C	at the closest mark to the top of the liquid
D	at the top of the plastic safety ring

Answer: B. at the bottom of the meniscus curve

To measure water in glass, you must look at the top surface at eye-level, and ascertain the location of the bottom of the meniscus (the curved surface at the top of the water). The meniscus forms because water molecules adhere to the sides of the glass, which is a slightly stronger force than their cohesion to each other. This leads to a U-shaped top of the liquid column, the bottom of which gives the most accurate volume measurement. (Other liquids have different forces, e.g. mercury in glass, which has a convex meniscus.) This is consistent only with answer (B).

15.	In an experiment measuring the growth of bacteria at different temperatures, what is the independent variable?
	(Average Rigor) (Competency 3.0)

A	number of bacteria
B	growth rate of bacteria
C	temperature
D	light intensity

Answer: C. temperature

To answer this question, recall that the independent variable in an experiment is the entity that is changed by the scientist, in order to observe the effects (the dependent variable(s)). In this experiment, temperature is changed in order to measure growth of bacteria, so (C) is the answer. Note that answer (A) is the dependent variable, and neither (B) nor (D) is directly relevant to the question.

16. **When designing a scientific experiment, a student considers all the factors that may influence the results. The process goal is to _____.**

   (Average Rigor) (Competency 3.0)

   A   recognize and manipulate independent variables
   B   recognize and record independent variables
   C   recognize and manipulate dependent variables
   D   recognize and record dependent variables

Answer: A.   recognize and manipulate independent variables

When a student designs a scientific experiment, s/he must decide what to measure, and what independent variables will play a role in the experiment. S/he must determine how to manipulate these independent variables to refine his/her procedure and to prepare for meaningful observations. Although s/he will eventually record dependent variables (D), this does not take place during the experimental design phase. Although the student will likely recognize and record the independent variables (B), this is not the process goal, but a helpful step in manipulating the variables. It is unlikely that the student will manipulate dependent variables directly in his/her experiment (C), or the data would be suspect. Thus, the answer is (A).

17.   **Which of the following is not an acceptable way for a student to acknowledge sources in a laboratory report?**
        (Rigorous) (Competency 3.0)

A   The student tells his/her teacher what sources s/he used to write the report.
B   The student uses footnotes in the text, with sources cited, but not in correct MLA format.
C   The student uses endnotes in the text, with sources cited, in correct MLA format.
D   The student attaches a separate bibliography, noting each use of sources.

Answer: A. The student tells his/her teacher what sources s/he used to write the report.

It may seem obvious, but students are often unaware that scientists need to cite all sources used. For the young adolescent, it is not always necessary to use official MLA format (though this should be taught at some point). Students may properly cite references in many ways, but these references must be in writing, with the original assignment. Therefore, the answer is (A).

18.   **Which of the following data sets is properly represented by a bar graph?**
        (Rigorous) (Competency 3.0)

A   Number of people choosing to buy cars, vs. Color of car bought.
B   Number of people choosing to buy cars, vs. Age of car customer.
C   Number of people choosing to buy cars, vs. Distance from car lot to customer home.
D   Number of people choosing to buy cars, vs. Time since last car purchase.

Answer: A. Number of people choosing to buy cars, vs. Color of car bought.

A bar graph should be used only for data sets in which the independent variable is noncontinuous (discrete), e.g. gender, color, etc. Any continuous independent variable (age, distance, time, etc.) should yield a scatter plot when the dependent variable is plotted. Therefore, the answer must be (A).

19.   **Chemicals should be stored**
        (Easy) (Competency 4.0)

A   in the principal's office.
B   in a dark room.
C   in an off-site research facility.
D   according to their reactivity with other substances.

Answer: D. According to their reactivity with other substances.

Chemicals should be stored with other chemicals of similar properties (e.g. acids with other acids), to reduce the potential for either hazardous reactions in the storeroom, or mistakes in reagent use. Certainly, chemicals should not be stored in anyone's office, and the light intensity of the room is not very important because light-sensitive chemicals are usually stored in dark containers. In fact, good lighting is desirable in a storeroom, so that labels can be read easily. Chemicals may be stored off-site, but that makes their use inconvenient. Therefore, the best answer is (D).

20.   **In which situation would a science teacher be legally liable?**
        (Average Rigor) (Competency 4.0)

A   The teacher leaves the classroom for a telephone call and a student slips and injures him/herself.
B   A student removes his/her goggles and gets acid in his/her eye.
C   A faulty gas line in the classroom causes a fire.
D   A student cuts him/herself with a dissection scalpel.

Answer: A. The teacher leaves the classroom for a telephone call and a student slips and injures him/herself.

Teachers are required to exercise a "reasonable duty of care" for their students. Accidents may happen (e.g. (D)), or students may make poor decisions (e.g. (B)), or facilities may break down (e.g. (C)). However, the teacher has the responsibility to be present and to do his/her best to create a safe and effective learning environment. Therefore, the answer is (A).

21. **Accepted procedures for preparing solutions should be made with**
   _____ .
   (Average Rigor) (Competency 4.0)

   A   alcohol
   B   hydrochloric acid
   C   distilled water
   D   tap water

Answer: C. distilled water

Alcohol and hydrochloric acid should never be used to make solutions unless instructed to do so. All solutions should be made with distilled water as tap water contains dissolved particles which may affect the results of an experiment. The correct answer is (C).

22. **Which is the most desirable tool to use to heat substances in a middle school laboratory?**
   (Average Rigor) (Competency 4.0)

   A   alcohol burner
   B   freestanding gas burner
   C   bunsen burner
   D   hot plate

Answer: D. Hot plate.

Due to safety considerations, the use of open flame should be minimized, so a hot plate is the best choice. Any kind of burner may be used with proper precautions, but it is difficult to maintain a completely safe middle school environment. Therefore, the best answer is (D).

23. **Electrophoresis uses electrical charges of molecules to separate them according to their:**
    (Rigorous) (Competency 4.0)

A   polarization.
B   size.
C   type.
D   shape.

Answer: B. size.

Electrophoresis uses electrical charges of molecules to separate them according to their size. The molecules, such as DNA or proteins, are pulled through a gel toward either the positive end of the gel box or the negative end of the gel box. DNA is negatively charged and moves toward the positive charge. Thus, the answer is (B).

24. **Experiments may be done with any of the following animals except**
    _____ .
    (Rigorous) (Competency 4.0)

A   birds
B   invertebrates
C   lower order life
D   frogs

Answer: A. birds

No dissections may be performed on living mammalian vertebrates or birds. Lower order life and invertebrates may be used. Biological experiments may be done with all animals except mammalian vertebrates or birds. Therefore the answer is (A).

25. In a science experiment, a student needs to repeatedly dispense very small measured amounts of liquid into a well-mixed solution. Which of the following is the best choice for his/her equipment to use? *(Rigorous) (Competency 4.0)*

A   pipette, stirring rod, beaker
B   burette with burette stand, stir-plate, beaker
C   volumetric flask, dropper, stirring rod
D   beaker, graduated cylinder, stir-plate

Answer: B.   burette with burette stand, stir-plate, beaker

The most accurate and convenient way to repeatedly dispense small measured amounts of liquid in the laboratory is with a burette, on a burette stand. To keep a solution well mixed, a magnetic stir-plate is the most sensible choice, and the solution will usually be mixed in a beaker. Although other combinations of materials could be used for this experiment, choice (B) is thus the simplest and best. Choices A and C rely on a stirring a rod, which requires the student to try to mix the solution by hand while adding the liquid. Answer D relies on a graduated cylinder to add the small amounts of liquid to the solution, and it is more difficult to add small quantities of a liquid with a graduated cylinder.

26. The elements in the modern periodic yable are arranged _____ .
(Easy) (Competency 5.0)

A   in numerical order by atomic number
B   randomly
C   in alphabetical order by chemical symbol
D   in numerical order by atomic mass

Answer: A.   in numerical order by atomic number
Although the first periodic tables were arranged by atomic mass, the modern table is arranged by atomic number, i.e., the number of protons in each element. (This allows the element list to be complete and unique.) The elements are not arranged either randomly or in alphabetical order. The answer to this question is therefore (A).

27.   **Which parts of an atom are located inside the nucleus?**
       (Easy) (Competency 5.0)

A   electrons and neutrons
B   protons and neutrons
C   protons only
D   neutrons only

Answer: B. protons and neutrons

Protons and neutrons are located in the nucleus, while electrons move around outside the nucleus. This is consistent only with answer (B).

28.   **What is specific gravity?**
       (Average Rigor) (Competency 5.0)

A   The mass of an object.
B   The ratio of the density of a substance to the density of water.
C   Density.
D   The pull of the earth's gravity on an object.

Answer: B. The ratio of the density of a substance to the density of water.

Mass is a measure of the amount of matter in an object. Density is the mass of a substance contained per unit of volume. Weight is the measure of the earth's pull of gravity on an object. The only option here is the ratio of the density of a substance to the density of water, answer (B).

29. The electrons in an atom that are used to form a chemical bond are
called _____.
(Average Rigor) (Competency 5.0)

A   outer shell electrons
B   excited electrons
C   valence electrons
D   reactive electrons

Answer: C.   valence electrons

Answers (A), (B), and (C) could all be used as terms to describe the electrons
involved in chemical bonding, depending on the situation. However only answer
(C) is the correct answer because it the specific name given to these electron
and not a description of the electrons.

30. The chemical equation for water formation is: $2H_2 + O_2 \rightarrow 2H_2O$.
Which of the following is an incorrect interpretation of this equation?
(Rigorous) (Competency 5.0)

A   Two moles of hydrogen gas and one mole of oxygen gas combine to
make two moles of water.
B   Two grams of hydrogen gas and one gram of oxygen gas combine to
make two grams of water.
C   Two molecules of hydrogen gas and one molecule of oxygen gas combine
to make two molecules of water.
D   Four atoms of hydrogen (combined as a diatomic gas) and two atoms of
oxygen (combined as a diatomic gas) combine to make two molecules of
water.

**Answer: B. Two grams of hydrogen gas and one gram of oxygen gas
combine to make two grams of water.**

In any chemical equation, the coefficients indicate the relative proportions of
molecules (or atoms), or of moles of molecules. They do not refer to mass,
because chemicals combine in repeatable combinations of molar ratios (i.e.,
number of moles), but vary in mass per mole of material. Therefore, the answer
must be the only choice that does not refer to numbers of particles, i.e., answer
(B), which refers to grams, a unit of mass.

31. **Which of the following is not a property of metalloids?**
(Rigorous) (Competency 5.0)

A   Metalloids are solids at standard temperature and pressure.
B   Metalloids can conduct electricity to a limited extent.
C   Metalloids are found in groups 13 through 17.
D   Metalloids all favor ionic bonding.

Answer: D. Metalloids all favor ionic bonding.

Metalloids are substances that have characteristics of both metals and nonmetals, including limited conduction of electricity and being in the solid phase at standard temperature and pressure. Metalloids are found in a 'stair-step' pattern from Boron in group 13 through Astatine in group 17. Some metalloids, e.g. Silicon, favor covalent bonding. Others, e.g. Astatine, can bond ionically. Therefore, the answer is (D). Recall that metals/nonmetals/metalloids are not strictly defined by the periodic table group, so their bonding is unlikely to be consistent with one another.

32. **Matter's phase (solid, liquid, or gas) is identified by its:**
(Easy) (Competency 6.0)

A   color and size.
B   shape and volume
C   size and volume
D   color and volume

Answer: B. shape and volume

Matter's phase is not dependent on nor has anything to do with its color, and the size of matter is not a necessary component in identifying its phase, eliminating answers A, C, and D. The correct answer is (B).

33. **When heat is added to most solids, they expand. Why is this the case?**
(Average Rigor) (Competency 6.0)

A   The molecules get bigger.
B   The faster molecular motion leads to greater distance between the molecules.
C   The molecules develop greater repelling electric forces.
D   The molecules form a more rigid structure.

Answer: B. The faster molecular motion leads to greater distance between the molecules.

The atomic theory of matter states that matter is made up of tiny, rapidly moving particles. These particles move more quickly when warmer, because temperature is a measure of average kinetic energy of the particles. Warmer molecules therefore move further away from each other, with enough energy to separate from each other more often and for greater distances. The individual molecules do not get bigger, by conservation of mass, eliminating answer (A). The molecules do not develop greater repelling electric forces, eliminating answer (C). Molecules form a more rigid structure when becoming colder and freezing (such as water). Water gives rise to an exception to heat expansion, so it is not relevant here, eliminating answer (D). Therefore, the answer is (B).

34. **Which of the following is most accurate?**
(Average Rigor) (Competency 6.0)

A   Mass is always constant; weight may vary by location.
B   Mass and weight are both always constant.
C   Weight is always constant; mass may vary by location.
D   Mass and weight may both vary by location.

Answer: A. Mass is always constant; weight may vary by location.

When considering situations exclusive of nuclear reactions, mass is constant. (Mass, the amount of matter in a system, is conserved.) Weight, on the other hand, is the force of gravity on an object, which is subject to change due to changes in the gravitational field and/or the location of the object. Thus, the best answer is (A).

35.   **The change in phase from liquid to gas is called:**
       (Rigorous) (Competency 6.0)

A   evaporation.
B   condensation.
C   vaporization.
D   boiling.

Answer: A. evaporation.

Condensation is the change in phase from a gas to a liquid; Vaporization is the conversion of matter to a vapor-—not all gases are vapors. Boiling is one method of inducing the change from a liquid to a gas; the process is evaporation. The answer is (A).

36.   **A seltzer tablet changing into bubbles is an example of:**
       (Rigorous) (Competency 6.0)

A   a physical change.
B   a chemical change.
C   conversion.
D   diffusion.

Answer: B. a chemical change.

A physical change is a change that does not produce a new substance. Conversion is usually used when discussing phase changes of matter, Diffusion occurs in aspects of a mixture when the concentration is equalized. A seltzer tablet changing into bubbles produces a new substance—gas—which is a characteristic of chemical changes.The answer is (B).

37.   **Energy is:**
        (Rigorous) (Competency 6.0)

A   The combination of power and work.
B   The ability to cause change in matter.
C   The transfer of power when force is applied to a body.
D   Physical force.

Answer: B. The ability to cause change in matter.

Power is the rate of doing work. The transfer of energy when force is applied to a body is work. The combination of power and work is not energy. Energy is simply the ability to cause change in matter. The answer is (B).

38.   **Catalysts assist reactions by _____ .**
        (Easy) (Competency 7.0)

A   lowering required activation energy
B   maintaining precise pH levels
C   keeping systems at equilibrium
D   changing the starting amounts of reactants

Answer: A.   lowering required activation energy.

Chemical reactions can be enhanced or accelerated by catalysts, which are present both with reactants and with products. They lower the required activation energy—so that less energy is necessary for the reaction to begin. Catalysts may require a well maintained pH to operate effectively, however, they do not do this themselves. A catalyst, by lowering activation energy, may change a reaction's equilibrium point, however, it does not maintain a system at equilibrium. The starting level of reactants is controlled separately from the addition of the catalyst. Thus the correct answer is (A).

39. **What is the best explanation for isomerization?**
   (Average Rigor) (Competency 7.0)

A   A chemical reaction in which a molecule changes shape, but no atoms are lost or gained.

B   A chemical reaction in which a molecule changes shape, but one or more atoms are lost.

C   A chemical reaction in which a molecule changes shape, but one or more atoms are gained.

D   A chemical reaction in which a compound is broke down into its constitute parts.

Answer: A. A chemical reaction in which a molecule changes shape, but no atoms are lost or gained.

Answers (B) or (D) would likely be decomposition reaction, Answer (C) is likely a combination or synthesis reaction. Answer (A) best explains isomerization in which a chemical rearranges its chemical bonds (these changing shape) without losing or adding any new atoms, although how those atoms are attached to each other will be different.

40. **Which of the following is a correct definition for "chemical equilibrium"?**
   (Average Rigor) (Competency 7.0)

A   Chemical equilibrium is when the forward and backward reaction rates are equal. The reaction may continue to proceed forward and backward.

B   Chemical equilibrium is when the forward and backward reaction rates are equal, and equal to zero. The reaction does not continue.

C   Chemical equilibrium is when there are equal quantities of reactants and products.

D   Chemical equilibrium is when acids and bases neutralize each other fully.

Answer: A. Chemical equilibrium is when the forward and backward reaction rates are equal. The reaction may continue to proceed forward and backward.

Chemical equilibrium is defined as when the quantities of reactants and products are at a 'steady state' and are no longer shifting, but the reaction may still proceed forward and backward. The rate of forward reaction must equal the rate of backward reaction. Note that there may or may not be equal amounts of chemicals, and that this is not restricted to a completed reaction or to an acid-base reaction. Therefore, the answer is (A).

41.   **Which of the following will not change in a chemical reaction?**
      (Average Rigor) (Competency 7.0)

A   number of moles of products
B   atomic number of one of the reactants
C   mass (in grams) of one of the reactants
D   rate of reaction

Answer: B.   atomic number of one of the reactants

Atomic number, i.e. the number of protons in a given element, is constant unless involved in a nuclear reaction. Meanwhile, the amounts (measured in moles (A) or in grams(C)) of reactants and products change over the course of a chemical reaction, and the rate of a chemical reaction (D) may change due to internal or external processes. Therefore, the answer is (B).

42.   **Carbon bonds with hydrogen by _____ .**
      (Rigorous) (Competency 7.0)

A   ionic bonding
B   non-polar covalent bonding
C   polar covalent bonding
D   strong nuclear force

Answer: C. polar covalent bonding

Each carbon atom contains four valence electrons, while each hydrogen atom contains one valence electron. A carbon atom can bond with one or more hydrogen atoms, such that two electrons are shared in each bond. This is covalent bonding, because the electrons are shared. (In ionic bonding, atoms must gain or lose electrons to form ions. The ions are then electrically attracted in oppositely-charged pairs.) Covalent bonds are always polar when between two non-identical atoms, so this bond must be polar. ("Polar" means that the electrons are shared unequally, forming a pair of partial charges, i.e. poles.) In any case, the strong nuclear force is not relevant to this problem. The answer to this question is therefore (C).

43. **Which reaction below is a decomposition reaction?**
    (Rigorous) (Competency 7.0)

A $HCl + NaOH \rightarrow NaCl + H_2O$
B $C + O_2 \rightarrow CO_2$
C $2H_2O \rightarrow 2H_2 + O_2$
D $CuSO_4 + Fe \rightarrow FeSO_4 + Cu$

Answer: C. $2H_2O \rightarrow 2H_2 + O_2$

To answer this question, recall that a decomposition reaction is one in which there are fewer reactants (on the left) than products (on the right). This is consistent only with answer (C). Meanwhile, note that answer (A) shows a double-replacement reaction (in which two sets of ions switch bonds), answer (B) shows a synthesis reaction (in which there are fewer products than reactants), and answer (D) shows a single-replacement reaction (in which one substance replaces another in its bond, but the other does not get a new bond).

44.   A cup of hot liquid and a cup of cold liquid are both sitting in a room at a temperature of 72 degrees Fahrenheit and 25% humidity. Both cups are thin plastic. Which of the following is a true statement? (Easy) (Competency 8.0)

A   There will be condensation on the outside of both cups.
B   There will be condensation on the outside of the hot liquid cup, but not on the cold liquid cup.
C   There will be condensation on the outside of the cold liquid cup, but not on the hot liquid cup.
D   There will not be condensation on the outside of either cup.

Answer: C. There will be condensation on the outside of the cold liquid cup, but not on the hot liquid cup.

Condensation forms on the outside of a cup when the contents of the cup are colder than the surrounding air, and the cup material is not a perfect insulator. This happens because the air surrounding the cup is cooled to a lower temperature than the ambient room, so it has a lower saturation point for water vapor. Although the humidity had been reasonable in the warmer air, when that air circulates near the colder region and cools, water condenses onto the cup's outside surface. This phenomenon is also visible when someone takes a hot shower, and the mirror gets foggy. The mirror surface is cooler than the ambient air, and provides a surface for water condensation. Furthermore, the same phenomenon is why defrosters on car windows send heat to the windows—the warmer window does not permit as much condensation. Therefore, the correct answer is (C).

45.   If the volume of a confined gas is increased, what happens to the pressure of the gas? You may assume that the gas behaves ideally, and that temperature and number of gas molecules remain constant. *(Average Rigor) (Competency 8.0)*

A   The pressure increases.
B   The pressure decreases.
C   The pressure stays the same.
D   There is not enough information given to answer this question.

Answer: B. The pressure decreases.

Because we are told that the gas behaves ideally, you may assume that it follows the ideal gas law, i.e., $PV = nRT$. This means that an increase in volume must be associated with a decrease in pressure (i.e., higher T means lower P), because we are also given that all the components of the right side of the equation remain constant. Therefore, the answer must be (B).

46.   What is necessary for ion diffusion to occur spontaneously? **(Diffusion relates to the movement of particles.)** *(Average Rigor) (Competency 8.0)*

A   carrier proteins
B   energy from an outside source
C   a concentration gradient
D   activation energy

Answer: C. a concentration gradient

Spontaneous diffusion occurs when random motion leads particles to increase entropy by equalizing concentrations. Particles tend to move into places of lower concentration. Therefore, a concentration gradient is required, and the answer is (C). No proteins (A), outside energy (B), or activation energy (which is also outside energy) (D) are required for this process.

47.   The relationships between pressure and temperature and between
        temperature and volume are examples of:
        (Average Rigor) (Competency 8.0)

A   indirect variations.
B   direct variations.
C   Boyle's law.
D   Charles' law.

Answer: B.   direct variations.

Charles' law deales strictly with the relationship between temperature and
volume. Boyle's Law deals with the relationship between pressure and volume.
These relationships are called direct variations because when one component
increases, the other decreases. The answer is (B).

48.   Which of the following is not true about phase change in matter?
        (Rigorous) (Competency 8.0)

A   Solid water and liquid ice can coexist at water's freezing point.
B   At 7 degrees Celsius, water is always in liquid phase.
C   Matter changes phase when enough energy is gained or lost.
D   Different phases of matter are characterized by differences in molecular
        motion.

Answer: B.   At 7 degrees Celsius, water is always in liquid phase.

According to the molecular theory of matter, molecular motion determines the
'phase' of the matter, and the energy in the matter determines the speed of
molecular motion. Solids have vibrating molecules that are in fixed relative
positions; liquids have faster molecular motion than their solid forms, and the
molecules may move more freely but must still be in contact with one another;
gases have even more energy and more molecular motion. (Other phases, such
as plasma, are yet more energetic.) At the 'freezing point' or 'boiling point' of a
substance, both relevant phases may be present. For instance, water at zero
degrees Celsius may be composed of some liquid and some solid, or all liquid, or
all solid. Pressure changes, in addition to temperature changes, can cause
phase changes. For example, nitrogen can be liquefied under high pressure,
even though its boiling temperature is very low. Therefore, the correct answer
must be (B). Water may be a liquid at that temperature, but it may also be a
solid, depending on ambient pressure.

49.   Pressure is measured in a unit called the pascal. One pascal is equal to one _____ of force pushing on one square meter of area.
        (Rigorous) (Competency 8.0)

A   joule
B   calorie
C   newton
D   erg

Answer: C. newton

A calorie is the amount of heat required to raise 1 g of water 1 degree (Celsius). An erg is a unit of energy that produces a velocity of 1 cm/sec in a mass of 1 gram. A joule is the amount of work done by one newton of force over one linear meter, while a pascal is equal to one newton of force pushing on one square meter of area. The answer is (C).

50.   All of the following are considered Newton's laws except for:
        (Easy) (Competency 9.0)

A   An object in motion will continue in motion unless acted upon by an outside force.
B   For every action force, there is an equal and opposite reaction force.
C   Nature abhors a vacuum.
D   Mass can be considered the ratio of force to acceleration.

Answer: C. Nature abhors a vacuum.

Newton's laws include his law of inertia (an object in motion (or at rest) will stay in motion (or at rest) until acted upon by an outside force) (A), his law that (Force)=(Mass)(Acceleration) (D), and his equal and opposite reaction force law (B). Therefore, the answer to this question is (C), because "Nature abhors a vacuum" is not one of these.

51. **A ball rolls down a smooth hill. You may ignore air resistance. Which of the following is a true statement?**
(Average Rigor) (Competency 9.0)

A   The ball has more energy at the start of its descent than just before it hits the bottom of the hill, because it is higher up at the beginning.

B   The ball has less energy at the start of its descent than just before it hits the bottom of the hill, because it is moving more quickly at the end.

C   The ball has the same energy throughout its descent, because positional energy is converted to energy of motion.

D   The ball has the same energy throughout its descent, because a single object (such as a ball) cannot gain or lose energy.

Answer: C. The ball has the same energy throughout its descent, because positional energy is converted to energy of motion.

The principle of conservation of energy states that (except in cases of nuclear reaction, when energy may be created or destroyed by conversion to mass), "Energy is neither created nor destroyed, but may be transformed." Answers (A) and (B) give you a hint in this question—it is true that the ball has more potential energy when it is higher, and that it has more kinetic energy when it is moving quickly at the bottom of its descent. However, the total sum of all kinds of energy in the ball remains constant, if we neglect 'losses' to heat/friction. Note that a single object can and does gain or lose energy when the energy is transferred to or from a different object. Conservation of energy applies to systems, not to individual objects unless they are isolated. Therefore, the answer must be (C).

52. **The force of gravity on earth causes all bodies in free fall to**

_____ .

(Average Rigor) (Competency 9.0)

A   fall at the same speed
B   accelerate at the same rate
C   reach the same terminal velocity
D   move in the same direction

Answer: B. accelerate at the same rate

Gravity causes approximately the same acceleration on all falling bodies close to earth's surface. (It is only "approximately" because there are very small variations in the strength of earth's gravitational field.) More massive bodies continue to accelerate at this rate for longer, before air resistance is great enough to cause terminal velocity, so answers (A) and (C) are eliminated. Bodies on different parts of the planet move in different directions (always toward the center of mass of earth), so answer (D) is eliminated. Thus, the answer is (B).

53. **Dynamics is the study of the relationship between:**
     (Rigorous) (Competency 9.0)

A   heat and energy.
B   heat and motion.
C   motion and work.
D   motion and force.

Answer: D. motion and force.

Dynamics is the study of the relationship between motion and the forces affecting motion. Force causes motion. The answer is (D).

54.   **Since ancient times, people have been entranced with bird flight. What is the key to bird flight?**
        (Rigorous) (Competency 9.0)

A   Bird wings are a particular shape and composition, that causes the air flow over the wing to travel faster than the air flow under the wing.
B   Birds flap their wings quickly enough to propel themselves.
C   Bird wings are a particular shape and composition, that causes the air flow under the wing to travel faster than the air flow over the wing.
D   Birds flapping of their wings creates a downward force that opposes gravity.

Answer: A.   Bird wings are a particular shape and composition, that causes the air flow over the wing to travel faster than the air flow under the wing.

Bird wings are composed of very light bones, and feathers. The shape of the feathers along the wing creates a curved upper surface and a flatter lower surface. The curved upper surface causes the air traveling over it to move faster that the air traveling under the wing, because the air has a greater distance to travel. Bernoulli developed the principle that when air traveling over two sides of a object are traveling at different speeds the a lifting force is created on the side that has the faster moving airflow. So the structure of the birds wing creates a lifting force. Without the proper structure no amount of flapping will bring about flight, if you doubt this just try flapping your own arms. Although flapping of wings does create some downward force this force is not sufficient to get a bird off the ground, without the lifting forces that the wing structure creates.   Thus, the answer is (A).

55.   **Newton's laws are taught in science classes because _____.**
       (Rigorous) (Competency 9.0)

A   they are the correct analysis of inertia, gravity, and forces
B   they are a close approximation to correct physics, for usual conditions on earth conditions
C   they accurately incorporate relativity into studies of forces
D   Newton was a well-respected scientist in his time

Answer: B. they are a close approximation to correct physics, for usual conditions on earth .

Although Newton's laws are often taught as fully correct for inertia, gravity, and forces, it is important to realize that Einstein's work (and that of others) has indicated that Newton's laws are reliable only at speeds much lower than that of light. This is reasonable, though, for most middle- and high-school applications. At speeds close to the speed of light, relativity considerations must be used. Therefore, the only correct answer is (B).

56.   **The transfer of heat by electromagnetic waves is called _____.**
       (Easy) (Competency 10.0)

A   conduction
B   convection
C   phase change
D   radiation

Answer: D. radiation

Heat transfer via electromagnetic waves (which can occur even in a vacuum) is called radiation. (Heat can also be transferred by direct contact (conduction), by fluid current (convection), and by matter changing phase, but these are not relevant here.) The answer to this question is therefore (D).

57.   A long silver bar has a temperature of 50 degrees Celsius at one end
      and 0 degrees Celsius at the other end. The bar will reach thermal
      equilibrium (barring outside influence) by the process of heat
      _____.
      (Average Rigor) (Competency 10.0)

A   conduction
B   convection
C   radiation
D   phase change

Answer: A. conduction

Heat conduction is the process of heat transfer via solid contact. The molecules
in a warmer region vibrate more rapidly, jostling neighboring molecules and
accelerating them. This is the dominant heat transfer process in a solid with no
outside influences. Recall, also, that convection is heat transfer by way of fluid
currents; radiation is heat transfer via electromagnetic waves; phase change can
account for heat transfer in the form of shifts in phase. The answer to this
question must therefore be (A).

58.   The law of conservation of energy states that:
      (Average Rigor) (Competency 10.0)

A   There must be the same number of products and reactants in any
      chemical equation.
B   Mass and energy can be interchanged.
C   Energy is neither created nor destroyed, but may change form.
D   One form energy must remain intact (or conserved) in all reactions.

Answer: C. Energy is neither created nor destroyed, but may change form.

Answer (C) is a summary of the law of conservation of energy (for non-nuclear
reactions). In other words, energy can be transformed into various forms such as
kinetic, potential, electric, or heat energy, but the total amount of energy remains
constant. Answer (A) is untrue, as demonstrated by many synthesis and
decomposition reactions. Answers (B) and (D) may be sensible, but they are not
relevant in this case. Therefore, the answer is (C).

59.   **When you step out of the shower, the floor feels colder on your feet than the bathmat. Which of the following is the correct explanation for this phenomenon?**
       (Rigorous) (Competency 10.0)

A   The floor is colder than the bathmat.
B   The bathmat being smaller that the floor quickly reaches equilibrium with your body temperature.
C   Heat is conducted more easily into the floor.
D   Water is absorbed from your feet into the bathmat so it doesn't evaporate as quickly as it does off the floor thus not cooling the bathmat as quickly.

Answer: C. Heat is conducted more easily into the floor.

When you step out of the shower and onto a surface, the surface is most likely at room temperature, regardless of its composition (eliminating answer (A)). The bathmat is likely a good insulator and is unlikely to reach equilibrium with your body temperature after a short exposure so answer (B) is incorrect. Although evaporation does have a cooling effect, it the short time it takes you to step from the bathmat to the floor, it is unlikely to have a significant effect on the floor temperature (eliminating answer (D)). Your feet feel cold when heat is transferred from them to the surface, which happens more easily on a hard floor than a soft bathmat. Therefore, the answer must be (C), i.e. heat is conducted more easily into the floor from your feet.

60.   **What is the best explanation of the term "latent heat"?**
       (Rigorous) (Competency 10.0)

A   The amount of heat it takes to change a solid to a liquid.
B   The amount of heat being radiated by an object.
C   The amount heat needed to change a substance to undergo a phase change.
D   The amount of heat it takes to change a liquid to a gas.

Answer: C. The amount heat needed to change a substance to undergo a phase change.

Answer (A) is a description of the term ' heat of fusion' and answer (D) is a description of the term 'heat of vaporization', both of which a specific examples of latent heat. Answer (C) includes both of these examples by using the term 'phase change' which includes the changes from solid to liquid, and liquid to gas. Answer (B) talks about at objects giving off heat without an accompanying change in state.

61.   **Sound waves are produced by _____ .**
       (Easy) (Competency 11.0)

A   pitch
B   noise
C   vibrations
D   sonar

Answer: C. vibrations

Sound waves are produced by a vibrating body. The vibrating object moves forward and compresses the air in front of it, then reverses direction so that the pressure on the air is lessened and expansion of the air molecules occurs. The vibrating air molecules move back and forth parallel to the direction of motion of the wave as they pass the energy from adjacent air molecules closer to the source to air molecules farther away from the source. Therefore, the answer is (C).

62.   **Sound can be transmitted in all of the following except _____ .**
       (Easy) (Competency 11.0)

A   air
B   water
C   a diamond
D   a vacuum

Answer: D. a vacuum

Sound, a longitudinal wave, is transmitted by vibrations of molecules. Therefore, it can be transmitted through any gas, liquid, or solid. However, it cannot be transmitted through a vacuum, because there are no particles present to vibrate and bump into their adjacent particles to transmit the waves. This is consistent only with answer (D). (It is interesting also to note that sound is actually faster in solids and liquids than in air.)

63. **The Doppler effect is associated most closely with which property of waves?**
    (Average Rigor) (Competency 11.0)

    A   amplitude
    B   wavelength
    C   frequency
    D   intensity

Answer: C. Frequency.

The Doppler effect accounts for an apparent increase in frequency when a wave source moves toward a wave receiver or apparent decrease in frequency when a wave source moves away from a wave receiver. (Note that the receiver could also be moving toward or away from the source.) As the wave fronts are released, motion toward the receiver mimics more frequent wave fronts, while motion away from the receiver mimics less frequent wave fronts. Meanwhile, the amplitude, wavelength, and intensity of the wave are not as relevant to this process (although moving closer to a wave source makes it seem more intense). The answer to this question is therefore (C).

64. **The speed of light is different in different materials. This is responsible for _____ .**
    (Average Rigor) (Competency 11.0)

    A   interference
    B   refraction
    C   reflection
    D   relativity

Answer: B. refraction

Refraction (B) is the bending of light because it hits a material at an angle wherein it has a different speed. (This is analogous to a cart rolling on a smooth road. If it hits a rough patch at an angle, the wheel on the rough patch slows down first, leading to a change in direction.) Interference (A) is when light waves interfere with each other to form brighter or dimmer patterns; reflection (C) is when light bounces off a surface; relativity (D) is a general topic related to light speed and its implications, but not specifically indicated here. Therefore, the answer is (B).

65. **A converging lens produces a real image _____.**
    (Rigorous) (Competency 11.0)

A   always
B   never
C   when the object is within one focal length of the lens
D   when the object is further than one focal length from the lens

Answer: D. when the object is further than one focal length from the lens

A converging lens produces a real image whenever the object is far enough from the lens (outside one focal length) so that the rays of light from the object can hit the lens and be focused into a real image on the other side of the lens. When the object is closer than one focal length from the lens, rays of light do not converge on the other side; they diverge. This means that only a virtual image can be formed, i.e. the theoretical place where those diverging rays would have converged if they had originated behind the object. Thus, the correct answer is (D).

66. **As a train approaches, the whistle sounds _____.**
    (Rigorous) (Competency 11.0)

A   higher, because it has a higher apparent frequency
B   lower, because it has a lower apparent frequency
C   higher, because it has a lower apparent frequency
D   lower, because it has a higher apparent frequency

Answer: A. higher, because it has a higher apparent frequency

By the Doppler effect, when a source of sound is moving toward an observer, the wave fronts are released closer together, i.e. with a greater apparent frequency. Higher frequency sounds are higher in pitch. This is consistent only with answer (A).

67. **The electromagnetic radiation with the longest wave length is/are**
    _____.
    (Easy) (Competency 12.0)

A   radio waves
B   red light
C   X-rays
D   ultraviolet light

Answer: A. radio waves

As one can see on a diagram of the electromagnetic spectrum, radio waves have longer wavelengths (and smaller frequencies) than visible light, which in turn has longer wavelengths than ultraviolet or X-ray radiation. If you did not remember this sequence, you might recall that wavelength is inversely proportional to frequency, and that radio waves are considered much less harmful (less energetic, i.e. lower frequency) than ultraviolet or X-ray radiation. The correct answer is therefore (A).

68. **In Ohm's law (I = V/R), the V represents:**
    (Average Rigor) (Competency 12.0)

A   current.
B   amperes.
C   potential difference.
D   resistance.

Answer: C. potential difference.

In Ohm's Law, I stands for current, which is measured in amperes. R stands for resistance, and V stands for potential difference. The answer is (C).

69.   **Which of the following is not a characteristic of all electrically charged objects?**
        (Average Rigor) (Competency 12.0)

A   Opposites attract.
B   Like repels like.
C   Charge is conserved.
D   A magnetic charge develops.

Answer: D. A magnetic charge develops.

All electrically charged objects share the following characteristics: Like charges repel one another. Opposite charges attract one another. Charge is conserved. While magnetic fields develop when charged objects move, this is not a ccharacteristic of all electrically charged objects. The answer is (D).

70.   **Resistance is measured in units called _____ .**
        (Average Rigor) (Competency 12.0)

A   watts
B   volts
C   ohms
D   current.

Answer: C. ohms

A watt is a unit of power. Potential difference is measured in a unit called the volt. Current is the number of electrons per second that flow past a point in a circuit. An ohm is the unit for resistance. The correct answer is (C).

71.   **A light bulb is connected in series with a rotating coil within a magnetic field. The brightness of the light may be increased by any of the following except:**
      (Rigorous) (Competency 12.0)

A   Rotating the coil more rapidly.
B   Using more loops in the coil.
C   Using a different color wire for the coil.
D   Using a stronger magnetic field.

Answer: C. Using a different color wire for the coil.

To answer this question, recall that the rotating coil in a magnetic field generates electric current, by Faraday's law. Faraday's Law states that the amount of electromotive force generated is proportional to the rate of change of magnetic flux through the loop. This increases if the coil is rotated more rapidly (A), if there are more loops (B), or if the magnetic field is stronger (D). Thus, the only answer to this question is (C).

72.   **A 10 ohm resistor and a 50 ohm resistor are connected in parallel. If the current in the 10 ohm resistor is 5 amperes, the current (in amperes) running through the 50 ohm resistor is**
      (Rigorous) (Competency 12.0)

A   1.
B   50.
C   25.
D   60.

Answer: A. 1.

To answer this question, use Ohm's law, which relates voltage to current and resistance: $V = IR$ where V is voltage; I is current; R is resistance. We also know that in a parallel circuit the voltage is the same across the branches.
Because we are given that in one branch, the current is 5 amperes and the resistance is 10 ohms, we deduce that the voltage in this circuit is their product, 50 volts. We then use $V = IR$ again, this time to find I in the second branch. Because V is 50 volts, and R is 50 ohm, we calculate that I is 1 ampere. This is consistent only with answer (A).

73. **Which is the correct sequence of insect development?**
(Easy) (Competency 13.0)

A   Egg, pupa, larva, adult.
B   Egg, larva, pupa, adult.
C   Egg, adult, larva, pupa.
D   Pupa, egg, larva, adult.

Answer: B. Egg, larva, pupa, adult.

An insect begins as an egg, hatches into a larva (e.g. caterpillar), forms a pupa (e.g. cocoon), and emerges as an adult (e.g. moth). Therefore, the answer is (B).

74. **Which part of a plant is responsible for transporting water.**
(Easy) (Competency 13.0)

A   phloem
B   xylem
C   stomata
D   cortex

Answer: B. xylem

The xylem transport a plant's food. A stomata is an opening on the underside of a leaf that allows for the passage of carbon dioxide, oxygen and water. The cortex is where a plant stores food. So the only answer is (B), which is where water is transported up the plant.

75. **Animals with a notochord or backbone are in the phylum ____.**
    (Average Rigor) (Competency 13.0)

A   arthropoda
B   chordata
C   mollusca
D   ammalia

Answer: B. chordata

The phylum arthropoda contains spiders and insects and the phylum mollusca contain snails and squid. Mammalia is a class in the phylum chordata. The answer is (B).

76. **What cell organelle contains the cell's stored food?**
    (Rigorous) (Competency 13.0)

A   vacuoles
B   Golgi apparatus
C   ribosomes
D   lysosomes

Answer: A. vacuoles

In a cell, the sub-parts are called organelles. Of these, the vacuoles hold stored food (and water and pigments). The Golgi apparatus sorts molecules from other parts of the cell; the ribosomes are sites of protein synthesis; the lysosomes contain digestive enzymes. This is consistent only with answer (A).

77.   Laboratory researchers have classified fungi as distinct from plants because the cell walls of fungi _____ .
        (Rigorous) (Competency 13.0)

A   contain chitin
B   contain yeast
C   are more solid
D   are less solid

Answer: A. contain chitin

Kingdom Fungi consists of organisms that are eukaryotic, multicellular, absorptive consumers. They have a chitin cell wall, which is the only universally present feature in fungi that is never present in plants. Thus, the answer is (A).

78.   A product of anaerobic respiration in animals is _____ .
        (Rigorous) (Competency 13.0)

A   carbon dioxide
B   lactic acid
C   oxygen
D   sodium chloride

Answer: B. lactic acid

In animals, anaerobic respiration (i.e. respiration without the presence of oxygen) generates lactic acid as a by-product. (Note that some anaerobic bacteria generate carbon dioxide from respiration of methane, and animals generate carbon dioxide in aerobic respiration.) Oxygen is not normally a by-product of respiration, though it is a product of photosynthesis, and sodium chloride is not strictly relevant in this question. Therefore, the answer must be (B). By the way, lactic acid is believed to cause muscle soreness after anaerobic weightlifting.

79. **In the law of dominance:**
    (Easy) (Competency 14.0)

A   Only one of the two possible alleles from each parent is passed on to the offspring.
B   Alleles sort independently of each other.
C   One trait may cover up the allele of the other trait.
D   Flowers have white alleles and purple alleles.

Answer: C. One trait may cover up the allele of the other trait.

Alleles sort independently of one another in the law of independent assortment; The law of segregation states that only one of the two possible alleles from each parent is passed on to the offspring. The color of flower alleles has nothing to do with any particular laws of inheritance. The answer is (C). The law of dominance says that in a pair of alleles, one trait may cover up the allele of the other trait.

80. **Which process(es) result(s) in a haploid chromosome number?**
    (Average Rigor) (Competency 14.0)

A   mitosis
B   meiosis
C   both mitosis and meiosis
D   neither mitosis nor meiosis

Answer: B. meiosis

Meiosis is the division of sex cells. The resulting chromosome number is half the number of parent cells, i.e., a "haploid chromosome number." Mitosis, however, is the division of other cells, in which the chromosome number is the same as the parent cell chromosome number. Therefore, the answer is (B).

81.   **A white flower is crossed with a red flower. Which of the following is a sign of incomplete dominance?**
      (Average Rigor) (Competency 14.0)

A   pink flowers
B   red flowers
C   white flowers
D   no flowers

Answer: A. pink flowers

Incomplete dominance means that neither the red nor the white gene is strong enough to suppress the other. Therefore both are expressed, leading in this case to the formation of pink flowers. Therefore, the answer is (A).

82.   **A child has type O blood. Her father has type A blood, and her mother has type B blood. What are the genotypes of the father and mother, respectively?**
      (Average Rigor) (Competency 14.0)

A   AO and BO
B   AA and AB
C   OO and BO
D   AO and BB

Answer: A. AO and BO

Because O blood is recessive, the child must have inherited two O's—one from each of her parents. Since her father has type A blood, his genotype must be AO; likewise her mother's blood must be BO. Therefore, only answer (A) can be correct.

83. **A monohybrid cross has four possible gene combinations. How many gene combinations are possible in a dihybrid cross?** *(Rigorous) (Competency 14.0)*

A   eight
B   sixteen
C   thirty-two
D   sixty-four

Answer: B. sixteen

In a monohybrid cross there are two possible contributions from each parent. For example if a parent's genotype is Gg, then the posibile contributions is either G or g. Since each parent has two possible ways to contribute, you then multiply the 2 possibilities form 1 parent times the 2 from the other parent to get 4 combinations in a monhybrid cross. In a dihybrid cross two genes are being considered for each parent, and so more combinations are possible. If a parent's genotype was GgRr, then the they might contribute either GR, Gr, gR, or gr to the child. WIth each parent contributing 4 possibilities, you end up with 16 gene combinations in the cross (4 from 1 parent times 4 possibilites from the other).

84. **An Arabic horse's purebred bloodline makes it a good example of:** *(Rigorous) (Competency 14.0)*

A   a homozygous animal.
B   a heterozygous animal.
C   codominance.
D   a polygenic character.

Answer: A. a homozygous animal.

A heterozygous animal is a hybrid. Codominance occurs when the genes form new phenotypes. A polygenic character is when many alleles code for one phenotype. A homozygous animal is a purebred, having two of the same genes present, as in a purebred horse breed. The answer is (A).

85.   A duck's webbed feet are examples of _____ .
        (Easy) (Competency 15.0)

A   mimicry
B   structural adaptation
C   protective resemblance
D   protective coloration

Answer: B. structural adaptation

Ducks (and other aquatic birds) have webbed feet, which makes them more efficient swimmers. This is most likely due to evolutionary patterns where webbed-footed-birds were more successful at feeding and reproducing, and eventually became the majority of aquatic birds. Because the structure of the duck adapted to its environment over generations, this is termed 'structural adaptation'. Mimicry, protective resemblance, and protective coloration refer to other evolutionary mechanisms for survival. The answer to this question is therefore (B).

86.   An animal choosing its mate because of attractive plumage or a strong mating call is an example of:
        (Average Rigor) (Competency 15.0)

A   sexual selection.
B   natural selection.
C   peer selection.
D   linkage.

Answer: A. sexual selection

The coming together of genes determines the makeup of the gene pool. Sexual selection, the act of choosing a mate, allows animals to have some choice in the breeding of its offspring. The answer is (A).

87. **Which of the following is not considered to be a cause of evolution?**
(Average Rigor) (Competency 15.0)

A   sexual reproduction
B   immigration
C   large populations
D   random mating

Answer: D. random mating

Evolution is caused by increase of the chances of variability in a population, which can be brought about by large populations, immigration, and simple sexual reproduction. Random mating actually decreases the chances of variability in a population, making the correct answer (D).

88. **Which of the following is not one of the principles of Darwin's theory of natural selection?**
(Average Rigor) (Competency 15.0)

A   More individuals are produced than will survive.
B   The Individuals in a certain species vary from generation to generation.
C   Only the fittest members of a species survive.
D   Some genes allow for better survival of an animal.

Answer: C. Only the fittest members of a species survive.

Answers (A), (B), and (D) were all specifically noted by Darwin in his theory. Answer (C) is often misquoted to represent this particular theory, but was not mentioned by Darwin himself.

89. **Which of the following is the best example of an explanation of the theory of evolution?**
     (Rigorous) (Competency 15.0)

A   Giraffes need to reach higher for leaves to eat, so their necks stretch. The giraffe babies are then born with longer necks. Eventually, there are more long-necked giraffes in the population.

B   Giraffes with longer necks are able to reach more leaves, so they eat more and have more babies than other giraffes. Eventually, there are more long-necked giraffes in the population.

C   Giraffes want to reach higher for leaves to eat, so they release enzymes into their bloodstream, which in turn causes fetal development of longer-necked giraffes. Eventually, there are more long-necked giraffes in the population.

D   Giraffes with long necks are more attractive to other giraffes, so they get the best mating partners and have more babies. Eventually, there are more long-necked giraffes in the population.

Answer: B. Giraffes with longer necks are able to reach more leaves, so they eat more and have more babies than other giraffes. Eventually, there are more long-necked giraffes in the population.

Although evolution is often misunderstood, it occurs via natural selection. Organisms with a life/reproductive advantage will produce more offspring. Over many generations, this changes the proportions of the population. In any case, it is impossible for a stretched neck (A) or a fervent desire (C) to result in a biologically mutated baby. Although there are traits that are naturally selected because of mate attractiveness and fitness (D), this is not the primary situation here, so answer (B) is the best choice.

90. **What is the principle driving force for evolution of antibiotic resistant bacteria?**
(Rigorous) (Competency 15.0)

A   mutation
B   reproduction method
C   population size
D   emigration

Answer: C. population size

Most bacteria reproduce asexually, so there is not a contribution to the variiabilty of the population. Thus answer (B) is not the driving force. Emigration, answer (D), or the act of moving away from an area, usually occurs after a bacteria has developed a resistance and not before. Mutation, answer (A), is a critical aspect in the evolution of antibiotic resistant bacteria, however useful mutations happen rarely and would be unlikely to develop a strain of antibiotic resistent bacteria. Only a large population size, answer (C), and the ability to quickly build that population would make it likely that a member of the colony would have devloped a useful mutation.

91. **What are the most significant and prevalent elements in the biosphere? (The biosphere contains all biomes)**
(Easy) (Competency 16.0)

A   carbon, hydrogen, oxygen, nitrogen, phosphorus
B   carbon, hydrogen, sodium, iron, calcium
C   carbon, oxygen, sulfur, manganese, iron
D   carbon, hydrogen, oxygen, nickel, sodium, nitrogen

Answer: A. carbon, hydrogen, oxygen, nitrogen, phosphorus

Organic matter (and life as we know it) is based on carbon atoms, bonded to hydrogen and oxygen. Nitrogen and phosphorus are the next most significant elements, followed by sulfur and then trace nutrients such as Iron, sodium, calcium, and others. Therefore, the answer is (A). If you know that the formula for any carbohydrate contains carbon, hydrogen, and oxygen, that will help you narrow the choices to (A) and (D) in any case.

92.   **Which of the following is the most accurate definition of a non renewable resource?**
       (Average Rigor) (Competency 16.0)

A   A nonrenewable resource is never replaced once used.

B   A nonrenewable resource is replaced on a timescale that is very long relative to human life spans.

C   A nonrenewable resource is a resource that can only be manufactured by humans.

D   A nonrenewable resource is a species that has already become extinct.

Answer: B.   A nonrenewable resource is replaced on a timescale that is very long relative to human life spans.

Renewable resources are those that are renewed, or replaced, in time for humans to use more of them. Examples include fast-growing plants, animals, or oxygen gas. (Note that while sunlight is often considered a renewable resource, it is actually a nonrenewable but extremely abundant resource.) Nonrenewable resources are those that renew themselves only on very long timescales, usually geologic timescales. Examples include minerals, metals, or fossil fuels. Therefore, the correct answer is (B).

93.   **A wrasse (fish) cleans the teeth of other fish by eating away plaque. This is an example of _____ between the fish.**
       (Average Rigor) (Competency 16.0)

A   parasitism
B   symbiosis (mutualism)
C   competition
D   predation

Answer: B. symbiosis (mutualism)

When both species benefit from their interaction in their habitat, this is called "symbiosis," or "mutualism." In this example, the wrasse benefits from having a source of food, and the other fish benefit by having healthier teeth. Note that 'parasitism' is when one species benefits at the expense of the other, 'competition' is when two species compete with one another for the same habitat or food, and 'predation' is when one species feeds on another. Therefore, the answer is (B).

94.   Which of the following terms does not describe a way that the human race has had a negative impact on the biosphere?
        (Rigorous) (Competency 16.0)

A   biological magnification
B   pollution
C   carrying capacity
D   simplifcation of the food web

Answer: C. carrying capacity

Most people recongize the harmful effects that pollution has caused, espically air pollution and global warming. The regular use of pesticides and herbicides introduces toxins in the food web. Biological magnification refers to how the concentration of these toxins increases the farther you move away from the source, so that animals at the top of the food chain, for example bald eagles, develop dangerous levels of toxins that may be responsible for declining birth rates in some species. Simplification of the food web has to do with small variety farming crops replacing large habitats, and thus shrinking or destroying some ecosystems. Carrying capacity, on the other hand, is simply a term that relates amount of life a certain habitat can sustain. It is term independent of human action, so the answer is (C). This is not to say that the number of humans is not having an impact. We are overpopulating the planet, and in so doing are moving past the carrying capacity of many habitats.

95.   Which one of the following biomes makes up the greatest percentage of the biosphere?
        (Rigorous) (Competency 16.0)

A   desert
B   tropical raine forest
C   marine
D   temperate deciduous forest

Answer: C. marine

All land biomes, which includes answers (A), (B), and (D) make up approximately 25% of the earth's surface, leaving the other 75% to the marine biome. Additionally the marine biome can range in depth from the air above the water, to several miles in depth. This combination makes answer (C) the correct answer.

96.   In commensalism:
       (Rigorous) (Competency 16.0)

A   Two species occupy a simlar place; one species benefits from the other
      and one species is harmed by the other.
B   Two species occupy a similar place and neither is harmed or benefits.
C   Two species occupy the similar place and both species benefit.
D   Two species occupy the same habitat and one preys upon the other.

Answer: B. Two species occupy a similar place and neither is harmed or benefits.

Answer (A) best describes a parastic relationship between the two species.
Mutualism (or symobiosis) describes the relationship that is seen in answer (C).
Anytime that a species can be considered prey, there is a predation relationship
between the species. This is answer (D). Leaving answer (B) as the description
of commensalism.

97.   When water falls to a cave floor and evaporates, it may deposit
       calcium carbonate. This process leads to the formation of which of
       the following?
       (Easy) (Competency 17.0)

A   stalactites
B   stalagmites
C   fault lines
D   sedimentary rocks

Answer: B. stalagmites

To answer this question, recall the trick to remember the kinds of crystals formed
in caves. Stalactites have a "T" in them, because they form hanging from the
ceiling (resembling a "T"). Stalagmites have an "M" in them, because
they make bumps on the floor (resembling an "M"). Note that fault lines and
sedimentary rocks are irrelevant to this question. Therefore, the answer must be
(B).

98.   Fossils are usually found in _____ rock.
       (Easy) (Competency 17.0)

A   igneous
B   sedimentary
C   metamorphic
D   cumulus

Answer: B. sedimentary

Fossils are formed by layers of dirt and sand settling around organisms, hardening, and taking an imprint of the organisms. When the organism decays, the hardened imprint is left behind. This is most likely to happen in rocks that form from layers of settling dirt and sand, i.e., sedimentary rock. Note that igneous rock is formed from molten rock from volcanoes (lava), while metamorphic rock can be formed from any rock under very high temperature and pressure changes. 'Cumulus' is a descriptor for clouds, not rocks. The best answer is therefore (B).

99.   Which of the following is the longest (largest) unit of geological time?
       (Average Rigor) (Competency 17.0)

A   solar year
B   epoch
C   period
D   era

Answer: D. era

Geological time is measured by many units, but the longest unit listed here (and indeed the longest used to describe the biological development of the planet) is the era. Eras are subdivided into periods, which are further divided into epochs. Therefore, the answer is (D).

100. **A contour line that has tiny comb-like lines along the inner edge indicates a?**
(Average Rigor) (Competency 17.0)

A   depression
B   mountain
C   valley
D   river

Answer: A.   depression

Contour lines are shown as closed circles in elevated areas and as lines with miniature perpendicular lined edges where depressions exist. These little lines are called hachure marks.

101. **Which of the following is a type of Igneous rocks?**
(Rigorous) (Competency 17.0)

A   quartz
B   granite
C   obsidian
D   all of the above

Answer: D. all of the above

Igneous rocks are formed from the crystallization of molten lava. Quartz takes the longest to crystallize (also slowest to cool) and tends to form large, distinct crystals.   Obsidian cools very quickly and forms a crystalline structure that appears to be smooth like glass. Granite is between Quartz and Obsidian, and is formed from a large number of small course crystals. Therefore, the answer is (D).

102. Lithification refers to one process to create _____.
     (Rigorous) (Competency 17.0)

A   metamorphic rocks
B   sedimentary rocks
C   igneous rocks
D   lithium oxide

Answer: B. sedimentary rocks

Lithification is the process of sediments coming together to form rocks, i.e., sedimentary rock formation. Metamorphic and igneous rocks are formed via other processes (heat and pressure or volcanoe, respectively). Lithium oxide shares a word root with 'lithification' but is otherwise unrelated to this question. Therefore, the answer must be (B).

103. Which of the following is the best explanation of the fundamental concept of uniformitarianism?
     (Rigorous) (Competency 17.0)

A   The types and varieties of life reveal a uniform progression over time.
B   The physical, chemical and biological laws that operate in the geologic past operate in the same way today.
C   Debris from catastrophic events (i.e. volcanoes, and meteorites) will be evenly distributed over the effected area.
D   The frequency and intensity of major geologic events will remain consistent over long periods of time.

Answer: B. The physical, chemical and biological laws that operate in the geologic past operate in the same way today.

While answers (A), (C), and (D) all could represent theories that have been proposed in geology, none of them is an accurate explanation of uniformitarianism. The general idea can be expressed, by the quote, "the present is the key to the past." The forces that we can observe today have been at work over most of earth's history.

104. **The salinity of ocean water is closest to _____ .**
(Easy) (Competency 18.0)

A   0.035 %
B   0.5 %
C   3.5 %
D   15 %

Answer: C. 3.5 %

Salinity, or concentration of dissolved salt, can be measured in mass ratio (i.e. mass of salt divided by mass of sea water). For Earth's oceans, the salinity is approximately 3.5 %, or 35 parts per thousand. Note that answers (A), (B) and (D) can be eliminated, because (A) and (B) are so dilute as to be hardly saline, while (D) is so concentrated that it would not support ocean life. Therefore, the answer is (C).

105. **The theory of "sea floor spreading" explains _____ .**
(Average Rigor) (Competency 18.0)

A   the shapes of the continents
B   how continents collide
C   how continents move apart
D   how continents sink to become part of the ocean floor

Answer: C. how continents move apart

In the theory of "sea floor spreading," the movement of the ocean floor causes continents to spread apart from one another. This occurs because crust plates split apart, and new material is added to the plate edges. This process pulls the continents apart, or may create new separations, and is believed to have caused the formation of the Atlantic Ocean. The answer is (C).

106. What is the most accurate description of the water cycle?
(Average Rigor) (Competency 18.0)

A   Rain comes from clouds, filling the ocean. The water then evaporates and becomes clouds again.

B   Water circulates from rivers into groundwater and back, while water vapor circulates in the atmosphere.

C   Water is conserved except for chemical or nuclear reactions, and any drop of water could circulate through clouds, rain, ground-water, and surface-water.

D   Weather systems cause chemical reactions to break water into its atoms.

Answer: C. Water is conserved except for chemical or nuclear reactions, and any drop of water could circulate through clouds, rain, ground-water, and surface-water.

All natural chemical cycles, including the water cycle, depend on the principle of conservation of mass. Any drop of water may circulate through the hydrologic system, ending up in a cloud, as rain, or as surface water or groundwater. Although answers (A) and (B) describe parts of the water cycle, the most comprehensive and correct
answer is (C).

107. What is the source for most of the United States' drinking water?
(Rigorous) (Competency 18.0)

A   desalinated ocean water
B   surface water (lakes, streams, mountain runoff)
C   rainfall into municipal reservoirs
D   groundwater

Answer: D. groundwater

Groundwater currently provides drinking water for 53% of the population of the United States. (Although groundwater is often less polluted than surface water, it can be contaminated and it is very hard to clean once it is polluted. If too much groundwater is used from one area, then the ground may sink or shift, or local salt water may intrude from ocean boundaries.) The other answer choices can be used for drinking water, but they are not the most widely used. Therefore, the answer is (D).

108. **Surface ocean currents are caused by which of the following?**
(Rigorous) (Competency 18.0)

A   temperature
B   density changes in water
C   wind
D   tidal forces

Answer: C. wind

A current is a large mass of continuously moving oceanic water. Surface ocean currents are mainly wind-driven and occur in all of the world's oceans (example: the Gulf Stream). This is in contrast to deep ocean currents which are driven by changes in density. Surface ocean currents are classified by temperature. Tidal forces cause changes in ocean level, however, they do not affect surface currents.

109. **Mount Kilauea on the island of Hawaii, is a very active volcano that has continuous lava flow into the ocean near it. What is the name of the type of shoreline created at the point where the lava flows meet the water?**
(Rigorous) (Competency 18.0)

A   stacking
B   submerged
C   developing
D   emergent

Answer: D. emergent

Answers (A) and (C) are not technical names for types of shorelines, although a stacked shoreline occurs when an island is worn down to rocks. In this case the lava is building on previously deposited lava and although the lava itself is submerging under the water to develop the shoreline the over all effect is the raising of the land out of the water. Thus the correct answer is (D).

110. **Which of the following instruments measures wind speed?**
     (Easy) (Competency 19.0)

A   barometer
B   anemometer
C   thermometer
D   weather vane

Answer: B. anemometer

An anemometer is a device to measure wind speed, while a barometer measures pressure, a thermometer measures temperature, and a weather vane indicates wind direction. This is consistent only with answer (B).

111.   The calm point at the center of a storm such as a hurricane is often called the "eye" of the storm. This "eye" is caused by:
       (Average Rigor) (Competency 19.0)

A   centrepidal force
B   a high-pressure air mass
C   a low-pressure air mass
D   heavier precipitation in the area

Answer: C. a low-pressure air mass

A large, low-pressure system accompanied by heavy precipitation and strong winds is called a hurricane. Heavier precipitation is not considered a characteristic of the eye of the storm. Centrepidal force would cause the low-pressure eye to expand outwards, dispersing the storm. The answer is (C).

112.   Air masses moving toward or away form the Earth's surface is called
       (Average Rigor) (Competency 19.0)

A   wind
B   breeze
C   air currents
D   doldrums

Answer: C. air currents

The doldrums, answer (D), describe the air masses near the equator. A breeze, answer (B), is a term used to describe winds created by local tempature changes. Wind, (answer (A), occurs when air masses move across the surface of the planet. Air currents, answer (C), is the term used to the vertical movement of air masses.

113. **Which type of cloud is most likely to produce precipitation?**
     (Average Rigor) (Competency 19.0)

A   cirrocumulus
B   stratocumulus
C   cumulonimbus
D   cirrostratus

Answer: C. cumulonimbus

Cirrocumulus and cirrostratus clouds (answers (A) and (D)) occur at the highest levels of cloud formation, and are thin veil-like or small patches, respectively. Stratocumulus clouds (answer (B)), occur low in the atmosphere, and are usually large irregular shaped puffs with large amounts of blue sky. These are the clouds that are usually used when looking for shapes in clouds. Leaving cumulonimbus clouds (answer (C)) to be correct. These clouds are most often associated with thunderstorms, these large, puffy, clouds have smooth or flattened tops, and can produce heavy rain and thunder.

114. **Air moving northward from the horse latitudes produces a belt of winds called the**
     (Rigorous) (Competency 19.0)

A   prevailing westerlies
B   north westerlies
C   trade winds
D   prevailing easterlies

Answer: A. prevailing westerlies

The prevailing westerlies are the winds found in the middle latitudes between 30 and 60 degrees latitude. They blow from the high pressure area in the horse latitudes towards the poles.

115. **Which is a form of precipitation?**
(Rigorous) (Competency 19.0)

A   snow
B   frost
C   fog
D   all of the above

Answer: A. snow

Snow is a form of precipitation. Precipitation is the product of the condensation of atmospheric water vapor that falls to the earth's surface. It occurs when the atmosphere becomes saturated with water vapor and the water condenses and falls out of solution. Frost and fog do not qualify as precipitates.

116. **Which of the following units is not a measure of distance?**
(Easy) (Competency 20.0)

A   AU (astronomical unit).
B   light year.
C   parsec.
D   lunar year.

Answer: D. lunar year

Although the terminology is sometimes confusing, it is important to remember that a 'light year' (B) refers to the distance that light can travel in a year. Astronomical units (AU) (A) also measure distance, and one AU is the distance between the sun and the earth. Parsecs (C) also measure distance, and are used in astronomical measuremen—they are very large, and are usually used to measure interstellar distances. A lunar year, or any other kind of year for a planet or moon, is the time measure of that body's orbit. Therefore, the answer to this question is (D).

117. **A star's brightness is referred to as**
     (Average Rigor) (Competency 20.0)

A   magnitude
B   mass
C   apparent magnitude
D   Intensity

Answer: A. magnitude

Magnitude is a measure of a star's brightness. The brighter the object appears, the lower the number value of its magnitude. The apparent magnitude is how bright an observer perceives the object to be. Mass has to do with how much matter can be measured, not brightness. The term intensity is not defined in reference to stars.

118. **Which of the following is the best definition for "meteroid"?**
     (Average Rigor) (Competency 20.0)

A   A meteroid is material from outer space, that is composed of particles of rock and metal.
B   A meteroid is material from outer space, that has struck the earth's surface.
C   A meteroid is an element that has properties of both metals and nonmetals.
D   A meteroid is a very small unit of length measurement.

Answer: A. A meteroid is material from outer space, that is composed of particles of rock and metal.

Meteoroids are pieces of matter in space, composed of particles of rock and metal. If a meteoroid travels through the earth's atmosphere, friction causes burning and a "shooting star"—i.e. a meteor. If the meteor strikes the earth's surface, it is known as a meteorite, answer (B). Answer (C) refers to a 'metalloid' rather than a 'meteroid', and answer (D) is simply a misleading pun on 'meter'. Therefore, the answer is (A).

119. **A telescope that collects light by using a concave mirror and can produce small images is called a _____.**
(Average Rigor) (Competency 20.0)

A   radioactive telescope
B   reflecting telescope
C   refracting telescope
D   optical telescope

Answer: B. reflecting telescope

Reflecting telescopes are commonly used in laboratory settings. Images are produced via the reflection of waves off of a concave mirror. The larger the image produced the more likely it is to be imperfect. Refracting telscopes use lenses to bend light to focus the image. The term optical telescope can be used to describe both reflecting and refracting telescopes.

120. **What is the main difference between the "condensation hypothesis" and the "tidal hypothesis" for the origin of the solar system?** *(Rigorous) (Competency 20.0)*

A   The tidal hypothesis can be tested, but the condensation hypothesis cannot.

B   The tidal hypothesis proposes a near collision of two stars pulling on each other, but the condensation hypothesis proposes condensation of rotating clouds of dust and gas.

C   The tidal hypothesis explains how tides began on planets such as Earth, but the condensation hypothesis explains how water vapor became liquid on Earth.

D   The tidal hypothesis is based on Aristotelian physics, but the condensation hypothesis is based on Newtonian mechanics.

Answer: B. The tidal hypothesis proposes a near collision of two stars pulling on each other, but the condensation hypothesis proposes condensation of rotating clouds of dust and gas.

Most scientists believe the "condensation hypothesis," i.e., that the solar system began when rotating clouds of dust and gas condensed into the sun and planets. A minority opinion is the "tidal hypothesis," i.e. that the sun almost collided with a large star. The large star's gravitational field would have then pulled gases out of the sun; these gases are thought to have begun to orbit the sun and condense into planets. Because both of these hypotheses deal with ancient, unrepeatable events, neither can be tested, eliminating answer (A). Note that both 'tidal' and 'condensation' have additional meanings in physics, but those are not relevant here, eliminating answer (C). Both hypotheses are based on best guesses using modern physics, eliminating answer (D). Therefore, the answer is (B).

121. **The planet with true retrograde rotation is _____.**
(Rigorous) (Competency 20.0)

A   Pluto
B   Uranus
C   Venus
D   Saturn

Answer: C. Venus

Venus has an axial tilt of only 3 degrees and a very slow rotation. It spins in the direction opposite of its counterparts (which spin in the same direction as the Sun). Uranus is also tilted and orbits on its side. However, this is thought to be the consequence of an impact that left the previously prograde rotating planet tilted in such a manner.

XAMonline, INC. 21 Orient Ave. Melrose, MA 02176
Toll Free number 800-509-4128

TO ORDER Fax 781-662-9268 OR www.XAMonline.com

MASSACHUSETTS TEST FOR EDUCATOR LICENTURE
- MTEL - 2008

PO#   Store/School:

Address 1:

Address 2 (Ship to other):

City, State Zip

Credit card number_____-_____-_____-_____   expiration_____

EMAIL _____

PHONE _____ **FAX** _____

ISBN	TITLE	Qty	Retail	Total
978-1-58197-287-0	MTEL Communication and Literacy Skills 01			
978-1-58197-876-6	MTEL General Curriculum (formerly Elementary) 03			
978-1-58197-607-8	MTEL History 06 (Social Science)			
978-1-58197-283-2	MTEL English 07			
978-1-58197-349-5	MTEL Mathematics 09			
978-1-58197-593-2	MTEL General Science 10			
978-1-58197-684-7	MTEL Physics 11			
978-1-58197-883-4	MTEL Chemistry 12			
978-1-58197-687-8	MTEL Biology 13			
978-1-58197-683-0	MTEL Earth Science 14			
978-1-58197-676-2	MTEL Early Childhood 02			
978-1-58197-893-3	MTEL Visual Art Sample Test 17			
978-1-58197-8988	MTEL Political Science/ Political Philosophy 48			
978-1-58197-886-5	MTEL Physical Education 22			
978-1-58197-887-2	MTEL French Sample Test 26			
978-1-58197-888-9	MTEL Spanish 28			
978-1-58197-889-6	MTEL Middle School Mathematics 47			
978-1-58197-890-2	MTEL Middle School Humanities 50			
978-1-58197-891-9	MTEL Middle School Mathematics-Science 51			
978-1-58197-266-5	MTEL Foundations of Reading 90 (requirement all El. Ed)			
			SUBTOTAL	
			Ship	$8.25
			TOTAL	

Printed in the United States
126224LV00002B/45-46/P